again, as is done on page 47, that properly speaking the theory of probability is a mathematical theory, or rather a set of such theories, so contrived as to be applicable to a wide variety of physical situations, in physics, in the social sciences, in insurance, in games of chance, and in a large number of other fields. The point is that the validity of a given application depends on how closely the physical situation conforms to the demands of the theory. For example, in the applications to throwing coins and rolling dice, it must be assumed that the coins and dice are perfectly symmetrically constructed of homogeneous material. But no real coins or dice fully meet these requirements. We are therefore talking about idealized coins and dice, and the problem arises: how closely can we expect the imperfect coins and dice that we use to conform to the laws of chance derived from the abstract theory? The fact that dice show small discrepancies from these laws, if enough thousands of rolls are made, is shown by the classical experiments of Weldon, discussed on pages 232-33 and on page 239.

Perhaps the most important matter to which I must refer in this brief Preface is the revisions that are *not* made in this edition. In Chapter XVII, beginning on page 276, there is a discussion of the field known as operations research (or operational research, as it is called in Britain). It is defined there as the application of scientific method to operational *military* problems. As such it was developed in England* during the Battle of Britain, as indicated in the text, and this remained its meaning on both sides of the Atlantic throughout World War II. At the end of the war the scientists of both countries recognized its great potential for other than military uses (pages 317-18). To further such uses there was

---

*For background material on operational research and on the relationship of government to the scientific effort in Britain, both during and preceding World War II, the reader can consult C. P. Snow's interesting book: *Science and Government*, Harvard University Press, 1961.

formed in England an Operational Research Club, later changed to a Society. In the United States in 1948 the National Academy of Sciences—National Research Council formed a Committee on Operations Research, with the same purpose of furthering the non-military applications, while certain universities, notably Massachusetts Institute of Technology and Case Institute of Technology, organized courses in the new field. *Scientific American* and *Fortune* magazines were the first to feature articles on the subject. As a result of these and other endeavors, operations research (and somewhat later its sister discipline, management science) has achieved a growth and an acceptance far beyond the hopes of those of us who were involved in its expansion. There can be little doubt that a major reason for this happy circumstance is the fact that operations research, in its broadest sense, received the official stamp of approval of science and scientists, at a time in history when science was held in high popular esteem.

In the 1939 edition of this book the author, perhaps rather timidly, expressed the hope and the conviction that at some future time the application of scientific method to the operational problems of business and to statistically complex situations in general, would come into its own. In the edition of 1950 much more space was devoted to an analysis of the difficulties involved, as they appeared to the author at that time (1948). These pages have been left unchanged in the present edition (pages 314-23). The fact is that the expansion and the development of the field of operations research have been so great that to bring our remarks on the subject up to date would require a separate volume.

Although the subject matter of this book is the elementary theories of probability and statistics and some of their simpler applications, wide segments of operations research are so intimately related to these theories that it properly finds a place in these pages. The examples of military operations research given in Chapter XVII were necessarily taken from work done during World War II from which, for one reason or another, security restrictions had been lifted. As to non-military work, the bulk of the two final chapters, which appear here unchanged from the 1950 edition, is in fact devoted to such applications. The majority of the illustrative examples were based on the author's personal experience in the field. As implied on page 317, these examples were selected on the basis of their close relationship to the subject of this book.

At that time (1948) what was urgently needed was an appropriate and descriptive term for the application of scientific method to the broad field of operations involving men and machines. In addition to being too restrictive, the term "business research," as used in the text, is a poor one. For there has long been intensive research in business and industry directed toward improving products and discovering new ones, and in some cases even extending into the area of basic research, none of which comes under the heading of operations research. The term "science of management" is highly descriptive, provided that it is confined to the functions of management, as usually understood. But operations research is not so confined.

It is strange that the English-speaking world should use the two equivalent terms "operational research" and "operations research." Since the British developed the field, at least in its military bearing, and since they coined the expression "operational research," it would seem that the fault lies on this side of the Atlantic. Where and how this schism developed I have never been able to learn. I do know that when the Committee on Operations Research was formed in 1948, the proposal to adopt the British designation was rejected on the ground that the term "operations research" was already too entrenched in the military establishments of this country. And so the schism will have to be perpetuated. All this is perhaps a counterexample to the usual implications of the phrase: "What's in a name?"

I would like to express my appreciation of the manner in which Dover Publications, and in particular Mr. Hayward Cirker, have handled the production of this edition.

HORACE C. LEVINSON.

Kennebunk, Maine,
July, 1962.

FOREWORD ·

ORIGINATING as the mathematical tool of the gambler, the science of probability has become fundamental in the knowledge of the physicist, the biologist, the technologist, the industrialist, the businessman, and the philosopher. Probability and its offspring statistics are now recognized as basic in the business world; also, it is becoming increasingly apparent that many principles of theoretical science have their explanation in the laws of chance. More frequently than many persons realize, data obtained from large numbers of similar events form patterns of behavior that the well-trained statistician or probability expert of sound judgment can recognize. Thus it becomes possible to formulate intelligent policies in the business world and to construct experiments and make predictions in the field of science.

Unfortunately, ignorance of even the very elementary principles of probability and statistics is widespread. Moreover, many persons who must apply statistical concepts do so in a routine and mechanical manner with no understanding of the methods employed or of the significance of the results obtained. It follows as an inevitable consequence that much "bad statistics" is being promulgated; the public has been victimized in many instances by pseudo statisticians who draw unjustified conclusions from data.

Horace C. Levinson, the author of *The Science of Chance*, is trying to develop literacy in the language of probability and statistics. He has assumed that a person need not be a skilled mathematician in order to understand the simple laws on which the science of chance is based. He takes the reader by the hand and leads him through the fundamentals of the subject by analyzing problems in games of chance; the concepts thus developed are then applied to problems of broad concern in government and business. Throughout the book the treatment reveals the clarity of the author's understanding of his subject as well as his pleasant sense of humor. The nonmathematician will read the book with genuine understanding, and he will be surprised that a mathematical subject can be so entertaining; the serious student of probability and statistics will find that the book provides an invaluable supplement to textbooks written on the subject.

The Science of Chance is a revision of the author's earlier book entitled *Your Chance to Win*. The latter work has been out of print for several years, and the publishers have received numerous requests that the book be reprinted. A particularly urgent request came from Professor S. S. Wilks of Princeton University, who, after reviewing galley proof of the new edition, has written, "We have not yet found another book which discusses the concepts of elementary probability theory and their role in everyday affairs and statistics in as concise, forceful, and simple language as Levinson's does. The main core of our course consists of a technical treatment of elementary statistical methods, and we believe Levinson's book will make excellent supplementary reading material."

Dr. Levinson's successful approach to his subject comes in part from his acquaintance with the broad utilization of probability concepts; although his Doctor's degree was in the field of mathematics, and he has worked actively in such an abstract field as the theory of relativity, he has been in great demand as a consultant and adviser to business and industrial organizations.

The publishers and the editor of this book present it to the public with considerable pride, for they believe that a greater knowledge of the science of chance is of fundamental significance in modern life.

C. V. Newsom.

Albany, N.Y.,
December, 1949.

CONTENTS ·

PART ONE

Chance

Probability is, for men, the guide of life.

JOSEPH BUTLER'S *Analogy*

Among other peculiarities of the nineteenth century is this one, that by initiating the systematic collection of statistics it has made the quantitative study of social forces possible.

ALFRED NORTH WHITEHEAD

CHAPTER I

Chance, Luck and Statistics

ALMOST everyone likes to take a chance once in a while, especially if the odds are not too much against him and the penalty for losing not too severe. Some people find in gambling for high stakes the answer to a craving for excitement. The quick shifts from high nervous tension to the letdown that follows apparently satisfy some primitive urge originally associated with a sense of danger. Others seem to enjoy risking their very lives, and too often the lives of others, as witness the thousands slaughtered unnecessarily each year in automobile accidents.

But *everyone*, whether he likes it or not, is taking chances every day of his life. No matter how cautious he is, or to what lengths he may go, he is still taking chances. To avoid taking chances he must avoid bathtubs, stairs, elevators, power machines of every sort. He dare not travel, while staying at home is dangerous in the extreme, as the statistics of the National Safety Council amply demonstrate. He is really in quite a fix. And finally, any nourishment he takes may be loaded with deadly bacteria and viruses; his only alternative is the certainty of starving to death.

To escape from this dilemma it is necessary to realize that taking a chance is an essential part of everyday life. We live in a world of chance, and if we wish to live intelligently we must know how to take chances intelligently. To do so we must know and understand the laws of chance, for there *are* laws of chance, although their existence is by no means self-evident.

In their efforts to understand the complexities of the world about them men were led, early in their history, to a series of fantastic beliefs that we now call superstitions. Some have persisted to the present day, in spite of the fact that they are flatly contradicted by the findings of science, while science is accepted by almost everyone. You will occasionally run across an individual who maintains that he does not believe in science, but if you will observe his actions you will find them explainable only on the assumption that he believes firmly in the laws of science, even though not conscious of the fact. Among the superstitious ideas that have persisted in a well-nourished state is that of luck. According to this notion chance events do not take place impartially. They are influenced by a mysterious something called *luck* that is attached to every man, like his shadow, except that he can rid himself of the latter by going into a dark place, while the former follows him everywhere.

It is curious that this notion of luck should be so widespread, for modern man is a skeptic and a realist. He is not easily induced to believe in things that he cannot see or touch. He does not believe in elves and brownies, or in ghosts that walk at midnight to the accompaniment of clanking chains. He does not believe in haunted houses, but he may believe that he himself is haunted by a spirit named luck. If it is good luck that is doing the haunting, his affairs will prosper. But without warning the other brother, bad luck, may take over, and against this misfortune he must be on perpetual guard.

Such ideas of luck are in fact misinterpretations of the idea of chance, and we shall see them for what they are when we have carefully surveyed the theory of chance, in more technical language the theory of probability, and the light that it throws on the interpretation of accidents and coincidences. We shall see that strange things are predicted by the laws of chance, freak runs of luck and striking coincidences. We shall be, perhaps, more tolerant and understanding of the origins of the superstitious doctrine of luck and less so of its survival.

From chance to statistics is but a step—how small a step will become evident in the pages that follow. Statistics is a highly useful and thoroughly practical subject, but it has its peculiar side as well. Even the word *statistics* is unusual. As a plural noun it means collections or sets of facts that are related, such as wheat yields per acre or the number of births per year. In practice it is customary to restrict its meaning to facts that are numerical, or can in some way be related to numbers. As a singular, collective noun statistics means the science of collecting or selecting statistical facts, sorting and classifying them, and drawing from them whatever conclusions may lie buried among them. The facts, for example, might have to do with the incidence of a certain disease. The conclusion might be that the disease thrives only when sanitation is poor. The singular form *statistic* is sometimes used to mean a single statistical fact. To avoid confusion it is important to keep these distinctions in mind. There is also the word *statistician*, which some people find difficult to pronounce. This word used to mean an expert, in other words a technically trained, highly competent person with specialized knowledge of the field of statistics. By long misuse it has degenerated to the point that it now means anyone, expert or rank amateur, who is in any way associated with statistics. It therefore has about it a flavor of disrepute, and it would be desirable to eliminate it from our vocabulary, and perhaps the whole statistics word family as well, if intelligible replacements were at hand. In their absence we shall bravely carry on, with the hope that the reader will not associate false meanings with such words.

One of the outstanding developments of our century, along with new techniques for destroying more people in less time, is the rapid expansion and development of the science of statistics. This science originated some hundred years ago in collections of facts bearing on the state, as the word implies. Exports, imports, births, deaths, economic trends, and the like, are all of interest to the state, and in the effort to learn from such facts the science of statistics gradually took shape. At last it became

· 5

clear that science had a powerful method for dealing with complicated patterns of facts, a method, furthermore, that works better the more facts there are and, in a sense, the more complicated the pattern. Here was the ideal working technique for all those subjects that study large masses of individuals, such as economics, sociology, government, insurance, even history.

For a long time, however, the older and more aristocratic sciences, like astronomy and physics, had little use for the new science of statistics. Radioactivity was not yet discovered; the foundations of physical science seemed secure and untroubled; the notion of a statistical law of nature would have been an absurdity to the scientists of the nineteenth century. Statistics was of use to them only as an artificial technique to handle situations too complicated to be dealt with in a straightforward, fundamental way.

Today all this is changed. Whole sections of astronomy and physics are full of statistics, and they are full of it not because they like statistics, but because it appears that nature does. Ordinary matter is made of molecules, and molecules are made of atoms, and atoms are made of electrons, protons, neutrons, and other less popularized particles. Years of observation and study have led to the view that these basic particles, which make up the world as we know it, follow statistical laws. This means that we can predict how very large crowds of them will act, but not what they will do individually. Those parts of physics that deal with the behavior of the electron are largely statistical in character. The same is true of the study of the nucleus of the atom. The theory of the atomic bomb, for example, is essentially statistical.

So statistics has found its way into all the sciences. In the social sciences it stands today as almost the sole practical working tool. Its importance in government is therefore immense. It is also of great value to biology, and thus to medicine.

Apart from the sciences proper, statistics is of rapidly increasing importance in other fields, for example in modern warfare, in modern business, and in practical politics. All this adds up to the fact that some knowledge of statistics is valuable to most of us in the art of living. Some enthusiasts have even attempted to insinuate statistics into the fine arts where, if it is not like the proverbial bull in a china shop, it is at least like a cow in a parlor. The statistical method has, in fact, like all rational procedures, sharp limitations, and it is no service to the cause of sound statistics to push it beyond its sphere. We shall have much to say in later chapters of unsound applications of statistics.

The subject of statistics has the dubious distinction of being, perhaps, the most misunderstood of all subjects. We do not mean that it is the least known. In order to misunderstand a subject it is essential that one know something of it, or at least think that he does. For example, my personal knowledge of the extensive field of paleopathology is so slight that I can make no claim of misunderstanding, only that of ignorance. In the case of statistics, however, it is commonly believed that anyone who is expert in school arithmetic is technically equipped, and that he can be made into a statistician by a process analogous to conferring a title. Actually, modern statistics is a highly developed technical field which has made much use of the resources of advanced mathematics. It is not possible to be a professional statistician, in the proper sense of the word, without a good deal of competence in mathematics. It is not necessary, however, to be a professional statistician in order to know the difference between good and bad statistics, or to use soundly the methods that have been developed. To prove the truth of this statement is a major aim of this book.

Statistics is also one of the most maligned of subjects. It is often said that with statistics you can prove anything, if you really try. There is a good deal of truth in this, at least if, in a spirit of congeniality, one drops the distinction between sound and unsound statistics. Unfortunately it is a fact that the majority of the statistical conclusions that reach the public are not based on sound statistics and are frequently biased. Unless they have to do with a simple alternative, like "guilty" or "not guilty," the chance of their being correct is small. This situation is a direct consequence of the common belief that a statistician does not need technical training and can be remedied only when this belief is discarded. If our leading newspapers published reports on technical medical subjects from nonmedical sources, there would be a parallel situation, and wise readers would skip the reports. If the wise reader followed the same procedure with respect to the statistical reports, he would have to skip a great deal. His alternative is a skeptical approach plus a cultivated ability to distinguish good statistics from bad.

In its rise to fame and achievement the theory of statistics has not stood alone. It was preceded by and has drawn much sustenance from another mathematical discipline, the theory of chance or probability. The latter theory, as we shall see, was born from problems that arose in connection with gambling and had reached a relative maturity while statistics was in diapers. It is not a mere historical accident that the theory of probability developed first. Statistical theory could not have gone very far without the fertile ideas provided by the other theory, ideas that are fundamental in the study of the collections of facts that are the central theme of statistics. One might say that probability and statistics look at the same situations through opposite ends of the telescope—probability through the eye end, that enlarges the image, statistics through the object end, that shrinks the image. Logically, the theory of probability comes first. It is not possible to understand statistics without it.

Before discussing the theory of statistics, it is therefore essential to explore the theory of probability as fully as we can. In so doing we shall in the main follow its historical development, beginning with simple problems in dice rolling and coin tossing, and later taking up its application to more complicated games like roulette, poker, and bridge. Such games provide an admirable approach to the central ideas of the theory of chance; here the laws of chance can be seen at work and fully grasped with a minimum of effort. It is hoped that enough different games are discussed to include at least one with which each reader is familiar, as familiarity with the game greatly increases its illustrative value. We shall also stop on our way to examine the role of chance in everyday life and in a wide range of subjects, roughly paralleling those mentioned in connection with statistics. For the two subjects are very closely interlocked. We shall also stop to examine some of the ideas for which chance is responsible—the notion of luck, good and bad, the many fallacies that arise in thinking about chance events, and the superstitions that have grown from such thinking.

The whole of Part I will be devoted to this exploration of chance and probability. The reader will encounter here the majority of the basic ideas essential to an intelligent approach to statistics, including all the more difficult or more subtle concepts. The new ideas introduced in connection with statistics, like frequency tables, various types of averages, and correlation, are more concrete and are more easily grasped.

It should not be inferred from what has been said that the sole purpose of Part I is to make possible Part II, which is devoted to statistics. The theory of probability is, in its own right, well worth careful study by all of us, as it enters our lives in hundreds of ways. Most important of all, perhaps, and not widely appreciated, is the intimate relation between this theory and the rules of sound thinking on practical matters. For we live in a world of uncertainties. The fact of tomorrow is the possibility of today. Probability is the rating we give to a possibility as a potential fact. It is unsound thinking to pick your favorite possibility and to proceed as though there were no others, even if your favorite leads the field. But it is not always an easy problem to know how to proceed soundly when several possibilities are taken into account. This problem is best illustrated in terms of gambling, and leads to the notion of expectation of profit or loss. We shall meet it in that connection.

The theory of probability not only enters our daily lives in many ways, but like its younger relative, statistics, it has penetrated and illuminated an extraordinarily wide group of subjects. Émile Borel, of the French Academy of Sciences, who has edited a comprehensive series of books on probability, has this to say: "It is thanks to the theory of probability that modern physicists explain the most hidden properties of energy and matter, that biologists succeed in penetrating the secret laws of heredity, thus permitting agriculturists to improve the stock of their animals and plants; in insurance, and in forecasts of every variety, the theory of probability is constantly used; . . . the theory of probability is of interest to artillerymen; it is of interest, also, not only to card and dice players, who were its godfathers, but to all men of action, heads of industries or heads of armies, whose success depends on decisions which in turn depend on two sorts of factors, the one known or calculable, the other uncertain and problematical; it is of interest to the statistician, the sociologist, the philosopher."

When we come to statistics itself we shall see how powerful an ally the theory of chance can be. This subject, without the intervention of the idea of chance, would be as dry as it is sometimes considered to be. It is when we look at statistical tables of figures as keys to the possibilities of the future that they come to life. If the statistics have to do, for instance, with the number of leaves on clover stems, you can infer the chance of finding a four-leaf clover among the next ten, or twenty, or one hundred specimens. If they have to do with baldness in men, you can compute the most probable number of bald heads that you will see at the next political meeting you attend, and checking up on the result may provide a pleasant means of passing the time during some of the speeches.

CHAPTER II

Gamblers and Scientists

THE idea of chance as something that can be treated numerically is comparatively new. For a long time men did not realize that they had the power to do so. By the famous and familiar allegory of the "Wheel of Fortune" they symbolized all the unknowable, mysterious, grotesque elements in their lives which combined to produce fate. What we now call chance was one of those elements. So potent was this concept of fate that it would have appeared almost impious to analyze it, at least for a long time, and a daring imagination was required to grasp the thought that numbers offered a key to the understanding of events which, superstitiously regarded, were utterly unpredictable.

That remarkable period in the history of science, the first half of the seventeenth century, produced the first attacks upon the problems of chance and probability. At this time, when the Renaissance was drawing to a close, a new curiosity about nature was sweeping the world, a curiosity destined to usher in the era of modern science. It extended even to the gamblers. A group of them, unable to answer their own questions about the fall of dice and other gaming problems, went to some of the leading scientists of the day. Among others, they approached the great Galileo.

The Italian gamblers met with an interested reception. Galileo, though occupied with a wide range of subjects, found these gambling problems worth careful study. Not only did he solve them all, but went on to write a short treatise on the game of dice. Science had begun its study of chance, which was to continue to the present day and lead to results that not even a Galileo could have foreseen. And it was science itself, rather than gambling, that was to receive the greatest benefits.

Not many years later history repeated itself, this time in France, and the first glimpses of a more general theory of chance were the result. Among the gamblers of the period there was one, a certain Chevalier de Méré, whose devotion to gaming was tempered by a scientific curiosity and insight which enabled him to formulate some very neat problems. Among his other virtues was his friendship for Blaise Pascal, and to him he brought his perplexing questions.

Pascal was a strange blend of mathematical genius, religious enthusiasm, and literary and philosophical ability of high order. The two streams of the mathematical and the mystical met in him, and their waters did not mingle. He was a battleground of opposing forces, which struggled for dominance throughout his rather brief life. In his early days mathematics reigned undisputed; his invention of the adding machine at the age of eighteen is but one of a series of accomplishments that gave him his high standing as a mathematician. Another is his solutions of de Méré's problems, for he was able to distill from them the basic ideas of a new branch of mathematics, often called the *theory of probability*.

Pascal created this novel intellectual instrument, but he was almost the first to violate the canons of his own creation. For the theory of probability teaches us above all to see in accidents and coincidences only the workings of natural law; it does away with the need of a supernatural symbolism in their interpretation. Pascal himself did not, apparently, accept the full implication of his own discovery. He retained a belief in the mysteriousness of certain accidents and occurrences. He was once driving in a suburb of Paris when his horses ran away, and he was saved from being thrown into the River Seine only by the traces breaking. This close call seemed to Pascal a special revelation warning him to abandon the world, and he lost no time in doing so. In order to fortify this resolution, from that day on he carried over his heart a piece of parchment on which he had written the details of the accident.

It may appear odd to modern readers that gamblers should have brought their problems to men of the importance of Galileo and Pascal. But a glimpse at the lives of one or two of the mathematicians of the period will throw light on this point. In retiring from the world early in life Pascal was not following the traditions then prevailing among the mathematical fraternity.

Today we are inclined to think of the mathematician as a voluntary exile, at least from the stormier phases of life, in spite of many exceptions, such as the late Paul Painlevé, three times Premier of France. But toward the close of the Renaissance, especially in the sixteenth century, the lives of many mathematicians read like pages from the autobiography of Benvenuto Cellini. There was Cardan, the Italian—physician, mathematician, astrologer, gambler. After obtaining his medical degree he was forbidden to practice, due to allegations that his birth was illegitimate. However, after a descent to the poorhouse he rebounded and finally obtained a renown second only to that of Vesalius. During lean years he supported himself with gambling. In his great book on algebra he published as his own a solution of cubic equations which he had obtained from a rival under pledge of secrecy, and it is called *Cardan's solution* to this day. One of his sons was executed for poisoning his wife; the other was a confirmed criminal who burglarized his father's house. In astrology one of his bizarre ideas was to cast the horoscope of Jesus Christ. For such follies he was arrested by the inquisition and ordered to publish no further books, but ended his life as a pensioner of the Pope. In addition to the hundred odd books he published he left behind more than one hundred in manuscript.*

*For further facts about Cardan, see Appendix I.

· 13

The name of Fermat is associated with that of Pascal in founding the new theory of chance. Fermat was a lawyer by profession; he never devoted himself wholly to mathematics. But his contributions to the subject are of so high an order that he is ranked as one of the great men of this field. Following the practice of the time, Pascal and Fermat corresponded extensively on scientific subjects, and in certain of his letters Pascal included the gambling problems of the Chevalier de Méré. Fermat responded with solutions that agreed with those of Pascal in results, but differed from them in method. Thus Fermat materially enriched the subject, but he published nothing concerning it. This was quite in accord with his usual practice, for he published very little on any subject, and his extensive and valuable researches in mathematics had to be dug up, after his death, from odd manuscripts and notes on the margins of the books he had been reading. Much of value was lost.

The rise of the theory of probability was rapid, as such things go. The first systematic treatise came some half century after the work of Pascal and Fermat and was due to James Bernoulli, the first of a family of Bernoullis almost as famous in mathematics as the Bach family in music.

We come next, after a lapse of one hundred years, to a great name in the history of science, that of Laplace. One of the inheritors of the genius of Newton, he spent most of his scientific life in wrestling with the mechanics of the heavens. Among his greatest works were his efforts to prove that the solar system is not subject to collapse, that the earth will not fall someday into the sun, that this same solar system evolved by orderly laws from a primitive gaseous nebula, and in constructing a monumental theory of probability. This work put the subject onto an entirely new plane and indicated clearly what its direction of growth was to be.

Laplace lived through the stormy period of the French Revolution and, in addition to his scientific labors, served in various administrative capacities. As to the value of these services opinions may differ, but one thing appears certain; his judgments of practical probabilities, in so far as his own advancement was concerned, were excellent.

The next great figure in the history of the theory of probability is the incomparable German mathematician, Gauss. Early in the nineteenth century Gauss undertook a study of the theory of measurements, which can be approached only by use of the idea of chance. His labors were so successful that he left this theory as perfectly developed as the mathematics of the period permitted. He evolved the famous law now known as the *Gauss Law of Errors,* which we shall meet again.

Gauss was one of the most versatile mathematical geniuses that ever lived. Everything that he touched seemed to turn to gold. In his admirable book, *Men of Mathematics,** E. T. Bell ranks Gauss, with Archimedes and Newton, as one of the three greatest mathematicians of all time.

Since the time of Gauss, the theory of probability has been developed and expanded by a host of workers, who have made use to the full of the immense resources of the mathematics of the past century.

If the theory of chance had grown up to be a black sheep in the scientific fold, we might hold the gamblers responsible for it and tell them that they should have kept their difficulties to themselves. But it is not a black sheep. It is a well-behaved member of the mathematical family and has rendered conspicuous service to many of that large group of subjects to which mathematics has been fruitfully applied. Among these subjects are not only many sciences, but many businesses as well, including insurance—one of the largest of all.

---

* The reader interested in knowing more of the men whose contributions to the theory of probability we have touched on will find here ample details of both their lives and their work.

So there is no question that science and business have contracted a very real debt to the gamblers. When science contracts a debt, it usually is able to repay it with interest by throwing a flood of light on the entire subject. In the present instance, the theory of probability has revolutionized the theory of games of chance and has pointed out how fallacious the theories of most gamblers are. Whether the gamblers have always appreciated this service is doubtful. At least it is fair to say that they should have.

CHAPTER III

The World of Superstition

In GAMES of chance there are bound to be runs of luck *for* and *against* the individual. This much is understood by everyone. No sane person sits down to a game of bridge with the expectation that all hands will be of about the same strength, nor does he expect the numbers in roulette to appear regularly in some order, so that, for instance, number 5 never would appear a second time until all the other numbers had had an opportunity to appear once. He expects these fluctuations due to chance —or luck, if he prefers to call it that—but he may see in them verifications of superstitions that he has previously acquired.

It is certainly true that around the gaming tables of the world there flourishes a particularly large crop of superstitions. Perhaps this is due primarily to the tension of gambling for high stakes. When the fall of a marble into hole number 4 instead of hole number 5, or the fall of an ace instead of a king, can catapult the player from poverty to riches, or vice versa, or when the turn of a card can mean a yacht or an estate on Long Island, it is not difficult to understand a lack of clear-cut, objective thinking. In the pages of this chapter we shall glance at some of the common superstitious beliefs in as detached and objective a frame of mind as possible, with emphasis on their relation to the laws of chance, and without the emotional strain under which a man who is gambling heavily must form his opinions.

Many years ago at Monte Carlo I had an opportunity to observe some of the curious human dramas at the famous casino.

After briefly trying out one of the "systems" given in a later chapter, I discovered a source of considerable amusement in watching the antics and eccentricities of some of the players. I recall one in particular, an elderly man who always entered the gaming rooms at the same hour and took a seat at a particular table. This man was well worth a second glance on account of a curious quality which showed in his expression. Aloof, tolerant, faraway, assured, he would enter the rooms with half a dozen notebooks and quantities of loose paper which he spread about him on the table, quite indifferent to the rights of his neighbors.

Standing behind him, I could see on these sheets a complicated pattern of numbers and hieroglyphics in columns. The job of keeping his entries up to the minute seemed to absorb him completely. Apparently he had no time to make bets. But this turned out to be a hasty opinion; the fact was that his calculations had not indicated that the moment for placing stakes had arrived. If I continued my observations long enough, perhaps for ten minutes, I would at last observe signs of activity. The latest calculations must have been successful, for counters would be extracted from an inner pocket and several bets placed with an air of precision. Sometimes he won, often he lost; in either case the mysterious calculations began again, and in either case his manner of quiet confidence continued.

The mystery of his play, however, was not unfathomable. Those queer-looking hieroglyphics turned out to be the signs of the zodiac. The elaborate calculations were to determine, according to the rules of astrology, whether the roulette table was *for* or *against* the player!

Quaint though the idea of looking to the stars for tips on the fall of a roulette marble may seem to some of us, it is a form of superstition which you may have encountered in some of your daily affairs without recognizing it! For instance: You are a businessman waiting at your office for a certain Mr. Jones to come in and sign a contract for a large order of your product.

He phones at last to say that he will be unable to get in today. But before the day is ended Mr. Jones unexpectedly runs across other arrangements that seem more advantageous to him, and your contract is never signed. That is all you will ordinarily discover about the way you came to lose a good order. The true story of what happened is this: On that particular morning Mr. Jones spilled the salt at the breakfast table. At once he made up his mind to be extremely cautious that day, and above all not to sign the contract. Then came the other arrangements. Mr. Jones believes that the salt spilling was a special warning sent by a kind providence to prevent him from entering into a disadvantageous contract. In fact, he says, if he had never believed in such signs before, a clear-cut case like this would have converted him at once.

Or you are sitting down to an evening of bridge when it appears that your partner has the firm conviction that dealing cards with red backs brings him bad luck. He therefore declines to cut for cards and deal. It is discreetly suggested that your opponents take the red deck, but one of them argues that if it brings bad luck to one person, it might very well bring it to another. He feels it more prudent to decline, and the game must be delayed until two neutral packs can be dug up, while you are wondering what the color of the backs can have to do with the hands dealt.

What is the significance of such actions, examples of which could be multiplied almost indefinitely? Why are these superstitious beliefs still so widespread and deep-rooted in a period like the present one, which is called the *scientific era?*

In the days when science was virtually the exclusive property of a few isolated individuals, it could not be expected that popular beliefs would be much influenced by it. Today the case is radically otherwise; as everyone knows, we have been fairly bombarded with scientific products and by-products. Home, office, street, and farm are teeming with them. No one can say that he has not been exposed more or less directly to the influence of science, or that he has not had a chance to "see the wheels go around." And the man in the street believes in science, or at least claims to. He has seen it work and, almost without realizing it, trusts his life daily to the accuracy of its predictions. When he sits down before his radio for an evening's entertainment he does not say, "I hardly think it will work tonight, as I just saw the moon over my left shoulder." If it does not work he examines the electrical connections, or looks for a faulty tube. It never occurs to him to connect the mishap with the number of black cats he has recently met. He knows very well that the difficulty is in the radio, and that if he cannot fix it the repairman will, regardless of the moon and the black cats. Yet this same man will sit down to a game of cards and find it perfectly natural to attempt to terminate a run of bad hands by taking three turns around his chair.

These superstitious beliefs are a heritage from a past, in which magic and the black arts, witchcraft, sorcery, and compacts with the devil were among the common beliefs of the people. Today many of them have passed away, especially those more obviously contradicted by the findings of modern science. It has been a long time since the last trial for witchcraft at Salem. The onetime fear of charms and curses, laid upon one person by another, is now a mark of a primitive stage of civilization. Curses, says Voltaire, will kill sheep, if mixed with a sufficient quantity of arsenic.

The majority of the superstitions that survive, though they are widespread, are of the more innocuous variety; belief in them seldom breeds the terror of the supernatural that characterizes inoculation by the more virulent species. Furthermore, there are many people who give them a wavering adherence, like that of the man who does not believe in ghosts but avoids cemeteries at midnight. Others find in them a pleasant opportunity for half-serious banter. Still others believe without being aware of it, like the woman who stated with pride that she had outgrown her superstition about the number 13. "Now," she said, "I frequently have thirteen at my dinners, for I have found out that thirteen is actually my lucky number." And it is a rare hostess, whether positively or negatively charged with superstition, or altogether devoid of it, who will serve thirteen people at dinner. In one of his speeches Chauncey Depew said, "I am not at all superstitious, but I would not sleep thirteen in a bed on a Friday night."

Although there is no denying that the modern form of superstition is a milder manifestation than previous ones, yet in a sense it is a much less excusable belief. For the increase in our knowledge of the natural world during the last three hundred years is immense. Where there was chaos and mystery there is now order and mystery. Science has not taken away the ultimate mystery, but it has added a good deal of order. It is more of an intellectual crime to see the world today through the dark glasses of superstition than it once was.

Superstition would be less prevalent if more people realized that there is only one set of natural laws, that things work only *one* way, not half a dozen ways. Nature abhors a contradiction. If a deck of cards really does take account of the fact that I dropped and broke my mirror this morning and distributes itself accordingly, why should I not expect my automobile to begin climbing telegraph poles under like circumstances? The one is a collection of fifty-two printed slips of paper, which operates according to the laws of chance, the other a steel mechanism that operates according to the laws of mechanics and thermodynamics. The distinction is less striking the more closely it is examined.

The fact is that if things really work in the way presupposed by these superstitious beliefs, then the reasoning of science is wrong from the bottom up, its experiments are delusions, and its successes merely happy accidents. As the successes of science number in the millions, this view of the world requires us to believe in millions of remarkable coincidences.

It is sometimes said that "the science of one age is the superstition of the next." In one sense nothing could be more misleading than this neatly turned remark, which implies that science and superstition differ only in that science is "true" or, at least, is accepted for the time being, while superstition is not. The real distinction is much more fundamental and lies in the manner of approaching a situation, not in the situation itself. Science represents the *rational* approach to the world. It observes and makes hypotheses, from which it reasons according to rules of logic. If conclusions get out of tune with observations, one or more of the hypotheses are thrown out or revised. If a mistake in the use of logical principles is committed, the defective parts are rejected, and a fresh start is made. Superstition, on the other hand, is an *irrational* approach to the world. It indicates as "cause" and "effect" pairs of events between which there is seriously insufficient observational or logical linkage. On this account it is, in a measure, protected from direct attack. There can be no question of pointing out errors in logic, where there is no logic. I walk under a ladder and that night my house burns down. Superstitious doctrine may point to the former as the "cause" of the latter, but fails to indicate any logical chain whatever between the two events. Walking under the ladder has put me under a "spell of bad luck," which has somehow communicated itself to the electrical wiring of my house, thus wearing away the insulation and causing the fire. The two events are thus connected; but not by any stretch of the imagination can the connection be called logical.

Superstition has its roots in egocentric thinking, in the individual's desire to regard himself as an important part of the scheme of things. Looking out from the meager prison of his own body at an overcomplex world which he can comprehend only in fragments, it is not astonishing that he tends to interpret what he sees in terms of himself. So it is with the gambler who watches the roulette wheel before playing, to find out "if the table is running with him." He has, unconsciously, raised himself to a position of considerable importance in the universe. Whether he wins or loses is by no means a matter of indifference to nature. Some days she is favorable to his winning, other days firmly opposed to it; he must keep a sharp watch and endeavor to confine his play to the former. He is like primitive man in his regard for the favor of gods and devils; but he brings no gifts to the goddess of chance; he sacrifices no heifers. He hopes, apparently, to take the goddess unawares and win her favor in the most economical manner. Perhaps he has less faith than his primitive brother.

There are only a few activities in which superstition comes even nearer to the surface than in the gambling of most card or dice players. Among them are the more hazardous occupations, such as the sea, wars, and mining. Whenever events over which the individual has no control can rob him of health or life, or can raise him suddenly to riches or reduce him to poverty, it is difficult for him to keep a sufficiently detached point of view to realize that the march of events takes as little account of him as we do of the insects that we unknowingly crush in walking across a lawn. An infantry captain who was decorated for bravery in action, and whom I met just after World War I, told me that he had never doubted his own safety, even in the tightest places, for he believed that he was natively "lucky," and that if this luck were about to desert him, he would infallibly recognize the fact. It was evident that this belief had given him great comfort and appreciably increased his courage. If superstitions of this nature, which are prevalent in all armies, always worked out as in the case of this infantry officer, they would furnish an example of the amount of good that a thoroughly bad idea can accomplish. But evidently they work out very differently. A soldier is wounded without the slightest premonition, or an acute sense of impending doom— caused perhaps by a case of indigestion—turns out to be merely a distressing and wholly unnecessary experience.

The idea that some people are inherently lucky in what they

undertake and others inherently unlucky is one of the superstitious doctrines most firmly intrenched in many minds. For
it is a matter of observation that chance sometimes favors individuals in streaks, called *runs of luck,* and from this fact it
is often erroneously concluded that some mysterious element
called *luck* resides in certain persons, and that another called
bad luck resides in others. The observation, as distinguished
from the inference, is quite correct and is perfectly in accord
with the laws of chance. In fact, this is the distinguishing feature of situations where chance enters. If no one ever had a
streak of good or bad luck in bridge hands, it would mean that
all of the hands dealt were of about the same strength, and the
element of chance, together with much of the interest, would
have disappeared from the game. It is correct and commonplace to say that a certain person *has been* lucky or unlucky in
some respect; it is sheer superstition to say that a person *will be*
lucky or unlucky in any respect. This indicates the correct
meaning of the word *luck.* When Goethe says that luck and
merit are traveling companions, he means not only that the
superior person is better able to take advantage of the twists
of fortune, but that he *appears* to have luck with him for that
very reason.

If we insist on interpreting the role of chance in the world
in terms of the experience of one individual, usually oneself,
we shall never succeed in understanding it. But we may in
that way come nearer to satisfying some of our latent vanities.
Apparently there are people who deliberately close their minds
to more enlightened views, preferring a flattering error to a
neutral truth. One is reminded of Chantecler in Rostand's
play. In his conversations with the glorious hen pheasant he
fights for his illusion; the sun could not rise without his crowing. He points with pride to the beauties of the dawn; the
world would be a sad place indeed if he should neglect his duty.
And when the hen pheasant suggests that he should do just
that, let a day pass without crowing, to prove to her that his

efforts really do cause the sun to rise, he sees that she does not believe in him and exclaims violently that if her suspicions are correct, he doesn't want to know it! Thus he preserves his illusion.

In order to see the workings of chance in perspective, it is best to leave the individual and turn to large groups of individuals, where experience multiplies very rapidly. By far the best illustrations come from games of chance. If you make a throw with two ordinary dice, the odds are 35 to 1 against your throwing the double six, and 1,679,615 to 1 against your making this throw four consecutive times. Suppose, though, that ten million people throw two dice simultaneously, four times in succession. Then it will be exceedingly likely that a few will throw four straight double sixes. Each individual who does so, however, finds it very astonishing that chance should have selected *him* as a favorite and is likely to look for a reason among the events of his own life, which have no more connection with his dice throw than do the spots on the sun.

Whenever anyone throws dice he is in effect taking part in a wholesale experiment, much larger, in fact, than the one imagined above, for the throws do not have to be made simultaneously—it does not make the slightest difference when they are made. When he has what appears to be a freakish series of throws, he should therefore keep in mind that if no one had such a series, it would be a gross violation of the laws of chance.

The same is true in real life, where such simple reckoning of future chances is rarely possible. It has required a million years of history, written and unwritten, for man to see beyond his own shadow and judge the external world according to standards appropriate to it, not those manufactured in his own image. Virtually all this immense stretch of time had passed before man reached the level of intelligence necessary to doubt his own supreme importance. Of the remaining few thousand years, modern science occupies only the last few hundred.

Through this period, as we saw in the previous chapter, the

theory of probability gradually grew out of the rolling of dice and the falling of cards. At the hands of the master mathematicians it took form, and its influence increased, until the foundation was laid for a rational view of chance events. It is our task to attempt to make clear the nature of this foundation.

The great value of games of chance in helping us to replace the superstitious view of the world by the rational one is due to their simplicity. In many of them it is easy to compute, using the rules of the theory of probability, just what the chances of the game are. This is equivalent to predicting what will happen if a large number of games is played; to compare these predictions with the results of experience is an easy matter, and one that should settle any doubts as to their correctness.

So, although games of chance are an excellent example of the validity of the laws of chance (to the discredit of the doctrines of superstition), devotees of games of chance are frequently superstitious in the extreme. These statements, put side by side, have an almost paradoxical flavor which disappears at once when the character of those addicted to play is taken into account. In the words of Anatole France, "Gamblers gamble as lovers love, as drunkards drink, inevitably, blindly, under the dictates of an irresistible force. There are beings devoted to gaming, just as there are beings devoted to love. I wonder who invented the tale of the two sailors possessed by the passion for gambling? They were shipwrecked, and only escaped death, after the most terrible adventures, by jumping onto the back of a whale. No sooner there than they pulled from their pockets their dice and diceboxes and began to play. There you have a story that is truer than truth. Every gambler is one of those sailors."

CHAPTER IV

Fallacies

IN THEIR private lives fiction writers, like other people, presumably have to obey the laws of nature. It must be as difficult for them to make water run uphill as it is for the rest of us, or to arrive at the last link of a chain of reasoning without bothering about the intermediate ones. But their pens are subject to no such laws. They have created paper characters with all sorts of astonishing powers of analysis, of divination, of control over others by hypnosis, of subduing the forces of nature by unheard-of inventions and have embedded them in equally astonishing situations. The resulting tales offer an avenue of escape, if only for an hour, from the reader's familiar scenes and the monotonous repetitions of ordinary life. But to forget the real world one must not stray too far from it. The success of the more fantastic yarns depends a good deal on the author's ability to bring these violations of natural law within the realm of the credible. This is particularly the case in certain tales of mystery and crime, where prodigious feats of detection are the focus of interest. In order to set the stage for the master detectives it is necessary to provide a never-ending sequence of interesting crimes in which the criminal is careful to leave just enough clues to provide two or three hundred pages of theory.

In these detective stories it is often the laws of chance rather than the victims which receive the most violent treatment. Not only are they grossly violated by the piling up of coincidence; they are as little respected in the detective's mental processes which in the end unravel the mystery. In his triumphant progress to the "logical" indication of the killer, the typical super-sleuth treats the process of reasoning as though it were an activity similar to crossing a stream by stepping from one stone to another. In real life the stages in the process of reasoning do not often lead to solid and immovable conclusions, but rather to conclusions that are merely more probable, or less so. When several such steps are linked together, the strength of the chain is determined by the laws of chance.

The character of Sherlock Holmes has probably appealed to more readers than any other in detective fiction. His spectacular and invariably accurate speculations, instead of leaving the reader with an annoying sense of their impossibility, produce the strong illusion that they could be imitated, if not duplicated, in real life, given a man of Holmes's acumen, powers of observation, and knowledge. In fact, there is a certain danger, much appreciated by Sherlock Holmes himself, that Dr. Watson, and the reader as well, will find his conclusions quite commonplace, once they have been allowed behind the scenes. One thing that conclusively prevents translating these fascinating and impressive performances from fiction into reality is the presence of that unpredictable element in things that gives rise to what we call *chance*. In real crime detection it is not generally true that the facts of the case admit of just one theory, the probability of which is overwhelming, but to several, each more or less probable. And it frequently happens that the odds are against each one of them, considered separately, just as in a horse race the odds are ordinarily against each horse. It is the "unlikely" and the "incredible" that happen daily. That Conan Doyle was aware of this fact is shown by Holmes's remark, after reconstructing the life history of Dr. Watson's brother from an examination of his watch: "I could only say what was the balance of probability. I did not at all expect to be so accurate." This must have been one of Holmes's more modest moods, for he can seldom be charged with such a lack of confidence in the accuracy of his conclusions!

If we turn back to Edgar Allan Poe, the originator of the modern detective story, we find a much keener appreciation of the principles of the theory of chance. This may be due to the fact that Poe was an enthusiastic student of mathematics in his school days and retained an interest in the subject throughout his life. In *The Mystery of Marie Roget,* with nothing more to go on than newspaper accounts, he accomplished the unique feat of giving, in narrative form, the correct solution to a genuine murder mystery which was unsolved at the time and which had long puzzled the New York police.

Mystery stories are not the sole offenders against the laws of chance. In *Peter Simple,* Captain Marryat tells of the midshipman who, during a naval engagement, was prudent enough to stick his head through the first hole in the side of the ship made by an enemy cannon ball, "as, by a calculation made by Professor Innman, the odds were 32,647 and some decimals to boot, that another ball would not come in at the same hole." This is a picturesque touch in a picturesque tale; like an epigram, it almost calls for immunity from attack, but it effectively illustrates the commonest and the deepest rooted of the fallacies connected with chance. Although easily elucidated, the error involved here is deep-rooted and has served as basis for much of the nonsense that has been written about gambling games. Let us take a closer look at the reasoning of this particular midshipman.

He was of course correct in believing that the chance that two balls would hit the ship at the same point is very small, but he was entirely mistaken in his belief that *after* a ball had hit, the chance that a second one would hit the same spot is smaller than the chance that it would hit any other spot, *designated in advance.* Before the engagement began, the betting odds were enormously against two balls' landing on the same spot, say a certain one of the portholes, but once half the "miracle" has been accomplished, the betting odds are immediately reduced to the odds that any indicated spot will not be hit.

To expect cannon balls to take account of where their predecessors have landed smacks of the supernatural, to say the least.

A similar practice was very common in the trench warfare of World War I. The idea was to take refuge in a newly made shell hole, because two shells never would fall on the same spot. Except in the case of systematic "sweeping," where the aim of the gun is changed periodically, from right to left, for example, there is no basis whatever for such a practice. As in the case of Marryat's sailor, it is not surprising that the tenants of the shell hole usually are safe, for there is much more room for shells to fall *outside* the hole than *inside* it. If it were possible to keep an exact record of the positions of men hit by shellfire, it would be found that in the long run those hiding in fresh shell holes fare no better than those concealed in older ones of the same size and conformation. It is perhaps just as well that this point cannot be checked by experience, as the soldier under fire cares little about the rules of thinking, and much about security, whether real or imaginary.

The same question comes up in other activities, in games of chance for instance, and here it is quite simple to answer it by experience. At the bridge table you consider it remarkable in the extreme if you pick up your hand and find thirteen cards of the same suit, and we agree at once that it *is* very remarkable, in the sense that it is very rare. But an equally remarkable and rare thing happens every time you pick up a hand. For at the moment that you reached for your cards the chance that your hand would contain the very cards it did contain was exactly equal to the chance that it would contain, say, thirteen spades. We attach a particular importance to the latter hand because, in the first place, it is so easy to describe and, in the second, because it has a particular significance in the game of bridge. If you find this fact astonishing, you may care to try the following practical experiment: Write down the names of thirteen cards, both suit and denomination, or better, specify one hundred such hands, and note the number of deals in future games required for you to hold one of these hands. There is only a very minute chance that you will ever hold one, but if enough people conducted the test, one or more would be almost certain to do so, sooner or later. The odds against holding a specified hand, like thirteen spades, are 635,013,559,599 to 1.

There is a similar situation in a lottery. Suppose that there are one million tickets at $1 apiece, each having the same chance to win. The holder of a single ticket has 1 chance in 1,000,000 to win the first prize, which will be an amount in six figures. From the standpoint of the ticket holder, it is almost a miracle to win the first prize, and only a little less miraculous to win any prize. But look at the matter from the point of view of the promoters of the lottery. For them it is an absolute certainty that there will be winners, and they see nothing miraculous about the names of the people who hold the winning tickets.

If we could look down on the run of everyday events the way promoters regard their lotteries, not as participants but as overseers, we should find these bizarre-looking things we call coincidences by no means so remarkable as they at first seem. We should continually see things taking place against which the odds were 1,000,000 to 1, or 100,000,000 to 1, but we should realize that if the event had turned out differently, the odds would also have been heavily against that particular outcome. Furthermore, we would see that except for the simplest situations, the odds are always enormously against what happens; we would see coincidences everywhere, and it would be difficult to single out any one of them as particularly remarkable. If the world were full of freaks, those in the circuses would attract little attention.

In *The Murders in the Rue Morgue*, Poe found occasion to express himself on the subject of coincidences and, incidentally, to air his enthusiasm for the theory of probability. During one of his long discourses Dupin says: "Coincidences, in general, are great stumbling-blocks in the way of that class of thinkers who have been educated to know nothing of the theory of probabilities—that theory to which the most glorious objects of human research are indebted for the most glorious of illustration."

If things are continually happening against which the odds are so heavy, what distinguishes an improbable or unbelievable coincidence from an ordinary or probable one? This is a most natural query, in view of the remarks that we have made, and one of the greatest importance in our effort to obtain a clear understanding of the workings of chance. When we encounter a series of such coincidences, perhaps in a laboratory experiment in science, it is essential to be able to say whether or not they can safely be attributed to pure chance alone. We are not yet prepared to answer this question fully, but we can at least sketch the method, leaving a more detailed discussion to later chapters.

Our first step must be to analyze the situation in which the coincidence, or set of coincidences, is embedded, in terms of the *laws of chance*. Now these laws are continually predicting, except in very simple situations, the occurrence of events with extremely small a priori probabilities, so that the fact that the odds against an event are very large is by no means significant, by itself. We must compare these odds with those against other events that could also occur. In our lottery the odds against winning first prize on a single ticket were 999,999 to 1, and this applied not only to the winning ticket, but to every other that was sold. So the fact that Henry Jones won is not at all remarkable; the laws of chance are indifferent to the name of the winner. But suppose that Henry had gone on to win two or three more similar lotteries. In that case the laws of chance would no longer remain silent. They would state emphatically that Henry's performance is unbelievable, on the hypothesis that chance alone was operating; for the odds against it are almost inconceivably large, compared with those against other events that could equally well have happened on the chance hypothesis.

The fallacy that permitted Marryat's midshipman to enjoy a somewhat false sense of security is nowhere more clearly exhibited than in certain gambling methods. Gamblers, especially those who have felt the impulse to record their theories in permanent form, have given it the imposing title of "the maturity of the chances." This means, in plain English, that in a game of chance you should favor bets on those chances which have appeared less frequently than others. In roulette, for example, if you notice that the number 5 has not appeared during the last one hundred or two hundred spins of the wheel, your chance to win, according to this method, will be greater if you back the 5 than if you back a number that has come up more frequently. This doctrine is based on the following reasoning: (a) In the long run the numbers come up equally often; (b) the 5 has appeared less often than the other numbers; therefore (c) the 5 should appear more frequently than the other numbers, until the "equilibrium" is restored.

Unless we succeed in making clear the fallacy in this reasoning, there is little use in discussing the theory of games of chance, or statistics and their applications to science and business. In particular, we shall need a clear understanding of the matter when we come to the question of "systems" in roulette. We shall therefore take a very simple example, familiar to everyone, where the point at issue stands out sharply:

Suppose that you play at heads or tails with a certain Mr. Smith, the amount bet at each toss of the coin being fixed at $1, and that Mr. Smith has allowed you to decide, before each toss, whether you will bet on heads or on tails. At a certain point in the game let us assume that heads has turned up ten consecutive times; how will you place your next bet, on heads or on tails? The doctrine of the "maturity of the chances" insistently advises you to bet on tails, "for," it says, "heads has come more than its share, at least in the last few tosses, and therefore there is more chance for tails, which will help to restore the balance."

The fallacy contained in this view cannot be more vividly expressed than in the phrase of Bertrand, the French mathematician, who remarks that the coin has neither memory nor consciousness. The coin is an inanimate object, and the reason that it is as likely to show heads as to show tails is its symmetrical construction, not any choice or fancy on its part. Now the construction of the coin, which alone determines its behavior, remains exactly the same, whether there have been ten straight heads, or ten straight tails, or one hundred straight tails, for that matter. On each toss there is always the same chance for heads as for tails. In your game with Mr. Smith it does not make the slightest difference to your prospects whether you decide to bet on heads or on tails after a run of ten consecutive heads.

"But," it may be objected, "it is certainly true that in the long run there ought to be as many heads as tails, and how can this be true unless there is a tendency for the one that has come up less frequently to catch up with the other one by appearing more frequently from that point on?" If we did not already know that the coin has no memory, and therefore cannot take account of its previous performance, this objection would seem, at first sight, to have some validity; in fact, it has deceived thousands of people into adopting this erroneous doctrine of "the maturity of the chances." The error is in the initial premise. It is not true that "in the long run there ought to be as many heads as tails." The correct statement is: In the long run we expect the *proportion* (or percentage) of heads and that of tails to be approximately equal. The *proportion* of heads means the number of heads divided by the total number of tosses; if the coin is tossed one thousand times, and there are 547 heads, the proportion of heads is 547 ÷ 1,000, or 547/1,000, while in this same series the proportion of tails is 453/1,000.

Now it is easy to see why in the long run the equality of the proportions of heads and of tails does not require that the one that has occurred less frequently tend to catch up with the other. Suppose that you have made ten thousand tosses and heads has appeared two hundred times more than tails— an unlikely but possible result. This difference, amounting to 1/50 of the number of tosses, appears to be very large, but if you continue to one million tosses, and the difference remains at two hundred, it now represents only 1/5,000 of the number of tosses, and the longer you continue, the smaller this proportion becomes.

This sort of thing is familiar to everyone in connection with people's ages. If you are older than your wife, even by five minutes, you can say, if you want to, "I was once twice as old as you were, and still earlier I was a million times as old." A difference of two years in age is enormous at the start of life, and almost negligible toward the end. When there is a deficit of two hundred tails in tossing a coin, it simply is not true that there is the slightest tendency for tails to appear more frequently than heads. If the tossing is continued, there is the same chance that the deficit be increased (in amount, *not* in proportion) as that it be decreased.

The best way to grasp this fundamental point—that in tossing a coin the chance of heads is exactly equal to the chance of tails, regardless of how the preceding tosses have come out— is to see that every other theory is absurd in itself and leads to absurd consequences. But there is another way, which is the last court of appeal, and this consists in making the experiment of tossing a perfectly symmetrical coin a large number of times, recording the results, and comparing them with those we are to expect, if it is true that heads is always as likely as tails. We should expect in the long run, for example, as many cases of ten heads followed by tails, as of ten heads followed by heads; that is to say, of a series of eleven consecutive heads. The exact basis for this comparison of theory and practice will be found later, in the chapter on the game of heads or tails.

The critical reader may ask at this point how we know that the coins used in such experiments are perfectly symmetrical. This is a legitimate question. If we know that the coins are symmetrical because the results agree with the theory, then we have not tested the laws of chance. We have *assumed* that the laws hold and have *proved* that the coins are symmetrical, while we wish to proceed in the reverse direction. The answer to this dilemma is that the symmetry or asymmetry of a coin is a physical question and must be established by physical means. Only after we are satisfied that the coin is of homogeneous material and symmetrical in shape can it be used to test the laws of chance. And it is obvious that ordinary coins, even assuming the material to be completely homogeneous, are disqualified by the fact that they have different designs on opposite sides. We must have a coin with the two sides identical, except for small spots of color, so that they can be distinguished.

Let us agree to pass such a coin. After all, it is hard to believe that a difference in color could have an appreciable effect, even in a long experiment. But what about dice? Ordinary dice, like ordinary coins, are disqualified by their different numbers of spots. In order to test the laws of chance we would need homogeneous cubes, each of the six sides painted with a different color, each color representing one of the numbers from one to six.

Whichever way we turn we find that similar precautions are necessary, if we wish to compare the predictions of the theory of probability with experience. In a roulette wheel, in addition to the question of perfect symmetry there is that of balance. In drawing balls from a bag, or cards from a deck, there is the question of shuffling. Like other experiments in science, those in probability require great skill.

It is customary in works on probability, and we have followed the practice in this book, to refer to experiments with coins and dice as though idealized, symmetrical coins and dice had been used. The difference in the results can be considerable. In Chapter XV we shall study an experiment in which ordinary dice were used. Twelve dice were thrown 4,096 times, the equivalent of 49,152 throws of a single die, and we shall point out that the laws of chance were *not* followed. This whole question is therefore a practical matter and must be kept in mind in the study of certain gambling games.

It is a curious, but not necessarily an astonishing fact that gamblers have a second doctrine that precisely contradicts that of the "maturity of the chances." According to this second principle you are advised to place your bets on the chances that have appeared *most* frequently. It would seem appropriate to call this maxim the "immaturity of the chances." From what I am able to make out of the accounts of this doctrine it appears that it should be followed when the maturity doctrine does not work. The decision as to when one of these theories ceases to work, and the other becomes valid, would seem rather difficult in practice.

It is remarkable that gamblers, for whom the theory of games of chance has a very real importance, are willing to trust to vague and theoretically unjustifiable theories, rather than to clear-cut experience. There is, however, a very important exception to this statement. Those gamblers who have, so to speak, elected to play from the other side of the table as proprietors of gambling establishments put their full faith in experience. They never fail so to establish the rules of play that the odds of the game are in their favor, and they trust for their profits to the uniform working of the laws of chance. These profits, therefore, are assured them, in the long run, provided that their working capital is sufficient to withstand the unfavorable runs of luck that are equally assured them by the laws of chance. We shall consider this interesting problem, of the relation of the gambler's available capital to his chance of success, in Chapter VIII. In studying this problem we are also studying a simplified form of the corresponding business problem, the relation of the working capital of a business to the chance of its failure.

CHAPTER V

The Grammar of Chance

ONE of the oldest extant ideas is the belief that there is an element of uncertainty or chance in the world, that in many of the events that take place on this planet there is something essentially unpredictable. The primitive peoples who held this belief personified chance as a host of gods and devils in whose wills and caprices all natural phenomena had their origin. The Greeks of Homer's day divided events into no less than three categories: those that the gods could alter at will, others that occurred according to immutable laws beyond the control of gods or men, and the events that respected neither gods nor laws. This last classification corresponds to what we mean by chance events.

To trace this idea of the fortuitous element in nature down through history would carry this book too far afield. But it was a concept which apparently was continually present in every subsequent period, from the time of the great Greek thinkers and philosophers through its later absorption into the doctrines of the church, and its relation there to the idea of Providence. As far as scientific philosophy is concerned, it began to sink into obscurity with the rise of the ideas of modern science, roughly three hundred years ago, particularly after the contributions of Descartes and his immediate followers.

This ancient notion of chance, after the time of Descartes, was replaced by the doctrine of determinism, based on exact natural law, of which Newton's law of gravitation became the model. Largely due to the far-reaching successes of this law, the doctrine of determinism remained firmly intrenched until the closing years of the last century, when the discovery of radioactivity, followed by the development of the quantum theory, laid the basis for a growing doubt as to its universal applicability. The door was thus opened for the return of the idea of a random, unpredictable element in the world. Today there is a definite revolt from the rigid form of determinism that ruled human thought for so long, both in the sciences and in philosophy. Yet if this idea does succeed in re-establishing itself, it will have a different form from that of its past appearance, for it will be inseparably tied to the notion of statistical law, with which there is always associated the idea of probability or chance.

In ordinary conversation we are continually using expressions that contain the idea of the probable, or the improbable. We say, "Not a chance in the world," or, "It is probably all over by now," or, "The thing is highly improbable." In fact, we use such expressions so frequently, and they form so intimate a part of our vocabularies, that it rarely occurs to us to give them a second thought; to ask, for instance, just what we mean by them.

When we pass from the domain of conversation to that of scientific discussion, we are at once struck by the altered and restricted use of many words. We do not, for example, ordinarily distinguish between the words *speed* and *velocity*. But in physics these words have distinct meanings; a velocity is a speed in a specified direction. To the uninitiated heat and cold are very different things, opposites, in fact. But in physics the word *cold* is altogether superfluous; there are only varying degrees of heat. Cold is a purely relative term, indicating an absence of heat, as gauged by any particular individual. If he lives in the tropics and encounters a temperature of 40° Fahrenheit, he will say that the air is chilly; an Eskimo inside the Arctic Circle would consider such a temperature warm. In the same way, the word *darkness* is superfluous; it means merely an absence of light. Similarly, in the theory of probability we shall have no need for the word *improbability*. Instead we shall consider improbable events as those for which the probability is less than one half. Whenever the probability is less than one half, it signifies that the "fair betting odds" are against the event in question. In this way our language is simplified and clarified; we speak of the probability of an event, regardless of whether it is probable or improbable, in the customary senses of these words.

Although we ordinarily express ideas involving the notion of chance in a very vague manner, when we turn to games and sports we are inclined to be a little more precise. There we adopt the language, if not the practice, of betting. We say, for instance, that "the odds are 2 to 1" in favor of one of the contestants.

What does this statement mean, precisely? If you say that the odds on throwing heads in tossing a coin are even, you mean that in a long series of tosses about half will be heads. Likewise, if you attempted to explain the statement that "the odds are 2 to 1 in favor of a certain contestant," you would have to say that the favored man will win two thirds of a long series of contests, the other, or others, one third. But this is a very unsatisfactory explanation. There are many forms of contest in which the element of chance is small, the element of skill large. Once a contestant has established his superior skill, the odds on future contests become overwhelmingly in his favor. The change in the odds is not due to any change in the contestants, but to a change in our knowledge of the contestants. In the same way the initial odds of 2 to 1 represent merely a rough estimate of the extent of our knowledge. When we turn to those games of chance in which the element of skill is entirely absent, the complete rules of the game take the place of these rough estimates. Everyone who knows them has the same amount of information, and the odds in favor of this or that result take on an impersonal flavor; they can be checked up by repeating the experience in question a large number of times.

Let us try to consider the chances of future events in a more precise manner. Suppose that the *fair* betting odds against a certain event are 3 to 1. This means that if repeated trials are made, the event in question must take place exactly once in every four trials, in the long run. Otherwise, one or other of the betters would win more and more money as time went on; this cannot happen if the odds are fair. We can say, then, that the chance or probability of the event is 1 in 4. We are leaving behind us vague ideas of future probabilities and introducing in their place evaluations in terms of numbers. As we have seen, it is this union of chance and number, this counting the spokes of the wheel of fortune, that first started the theory of probability along the right road.

Whenever we learn how to measure anything, our next step is to find out how to add, subtract, multiply, and so on, in terms of the new things. How do we add chances? Suppose that the chance of one event is 1 in 6, that of another 1 in 2, and that we have both a reason and a right to add the two chances. (We shall find out later the rules that justify the adding of chances.) The answer is very simple. Let us agree to write all chances in the form of fractions; the above become respectively $\frac{1}{6}$ and $\frac{1}{2}$, and to add them we follow the rules of arithmetic for adding fractions, getting for the sum $\frac{4}{6}$ or $\frac{2}{3}$. Similarly for subtraction and other operations, we merely follow the rules of arithmetic. From now on we shall read the statement "the chance is $\frac{1}{5}$" as "the chance is 1 in 5." Since $\frac{1}{5}$ is the same as 0.2, using decimals, we can also say "the chance is 0.2," which is to be read "the chance is 2 in 10, or 1 in 5."

Consider next one of a pair of dice, an ordinary die with six sides. If I ask what the odds are against throwing any one of the sides, say the six spot, you will reply that they are evidently 5 to 1, and your reply will be not only evident but, more important, correct. If throwing a die were our only problem, we might let the matter end there. In order to know how to deal with more complicated problems, however, we shall have to pick this evident statement to pieces and learn from it a rule that will work for all cases. If I ask you further why you say that the odds are 5 to 1 against the six spot, you will perhaps answer, "The die is symmetrically constructed, and if you throw it a large number of times there is no conceivable reason why one of its sides should turn up more frequently than another. Therefore each side will logically turn up on about one sixth of the total number of throws, so that the odds against any one of them, the six spot, for instance, are very nearly 5 to 1." Or you may say, "At each throw there are six possible results, corresponding to the six faces of the die, and since the latter is symmetrically constructed, they are all equally likely to happen. The odds against the six spot are 5 to 1."

Instead of saying that the odds are 5 to 1 *against* the six spot, we can as well say that the chance or probability of the six spot is 1 in 6, or $\frac{1}{6}$.

The first of these replies refers to the result of making a long series of throws. What it defines is therefore called a *statistical* probability. Notice the words *very nearly*.

The second reply refers to a single throw of the die, and therefore appears to have nothing to do with experience. What it defines is called an *a priori* probability.

Each of the probabilities just defined is equal to $\frac{1}{6}$. This is an example of a general law known as the *law of large numbers*, which tells us that these two probabilities, when both exist, are equal. More accurately, they become equal when the number of throws (or trials) is indefinitely increased. We shall therefore drop the Latin adjectives and speak simply of the probability.

It is the second of these forms that gives us the clue we are after. According to it, the chance or probability of throwing the six spot is simply the number of throws that give the 6 (one), divided by the total number of possible throws (six). What if we adopt this rule, restated in more general language, as the numerical *definition* of probability? This rule, then, will tell us what we mean by a probability, expressed as a number. We are always free to choose any meaning that we desire for a new term, provided only that we do not in any way contradict ourselves. In the present instance, we have already seen that this rule agrees with our preconceived notions about betting odds. Furthermore, the probabilities that we find by the use of this rule can be checked by experience.

Before stating the rule in general form, there is a small matter concerning a convention of language that we must agree upon. Suppose that we are discussing the probability of some event, say that of drawing an ace from an ordinary pack of fifty-two cards. On each draw either an ace is drawn or some card not an ace. To avoid clumsy circumlocutions we shall agree to refer to the first result (that of drawing an ace) as a *favorable case,* to the second result as an *unfavorable case.* A like convention will hold no matter what the event in question. Thus the possible cases are always the favorable cases plus the unfavorable cases.

So we come to the fundamental definition of a probability. We shall first introduce it formally, and then try to get at what it means on a more informal basis:

The probability of an event is defined as the number of cases favorable to the event, divided by the total number of possible cases, provided that the latter are equally likely to occur.

This practical rule tells us how to go about finding the probabilities or chances in a large variety of problems, in particular those relating to games of chance. The procedure it indicates is as follows: (*a*) Make a list of all the possible and equally likely cases (for the die this is a list of its six sides). (*b*) Pick from this list those cases which give the event in question (for the problem of the die which we have been considering, this second list contains only the "six spot"). (*c*) The probability of the event is the fraction whose numerator is the number of cases in the second list, and whose denominator is the number of cases in the first list.

To illustrate the use of this rule further by applying it to another simple problem with one die: What is the chance of throwing *either* a five spot or a six spot? First list the possible cases. With one die there are always six, the number of its sides, in other words. Next pick out the favorable cases, which are the five spot and the six spot; there are two. The probability we are looking for is therefore the fraction whose numerator is 2 and whose denominator is 6; it is $\frac{2}{6}$ or $\frac{1}{3}$.

Every probability that can be expressed in numbers takes the form of a fraction. We must notice at once that the value of this fraction can never be greater than 1; in other words, such a probability as $\frac{4}{3}$, say, does not exist. To see that it cannot, we need only express it in the form "the chance of the event is 4 in 3," a very absurd statement. Or, using the fundamental rule, we can say that a probability of $\frac{4}{3}$ means that of three possible cases four are favorable, which is outright nonsense. What if the probability of an event is $\frac{1}{1}$ or 1? This means that all the possible cases are favorable; in other words, the event is *certain* to happen. Similarly, if the probability is 0, it means that none of the possible cases, no matter how many there are, is favorable. The event in question is certain *not* to happen; it is *impossible*.

The probability of tossing *either* heads or tails with one coin is evidently $\frac{2}{2}$ or 1, for both the possible cases are favorable. This is an example of a certainty. The probability of throwing a total of 13, or of 1, with two dice is an example of an impossibility, or 0 probability.

Every probability is thus equal to a fraction whose numerator is not greater than its denominator, and we have enlarged the ordinary sense of the word *probability* to include both certainty and impossibility. It is perhaps unnecessary to add that in referring to fractions ordinary decimal fractions are included. A decimal fraction is merely a fraction whose denominator is a power of 10; thus, 0.359 means 359/1,000.

Before applying our fundamental rule to more complicated situations, where the various chances are not so evident, let us take a careful look at it, above all at the proviso that is tagged on at the end.

This proviso, if you recall, states that the definition applies only in case *all* the possible occurrences are equally likely. This means that unless we are in a position to make in advance a common-sense judgment of the sort made in the case of the die, which tells us that each of its sides is as likely to turn up as another, the rule will not work. We cannot compute the chances of the various possibilities; to evaluate them we must appeal to actual experience and the so-called law of large numbers.

This is the situation in most problems other than simple games of chance. In life insurance, for example, the fundamental mortality tables are derived from actual experience. In considering the chances of the death of any individual, there is no clear meaning to be attached to the expression "equally likely events." Or consider an advertising problem where millions of individual prospects are involved. Experience must teach us something of the basic probabilities before mathematical statistics can take a hand.

On the other hand, when we do succeed in analyzing a problem into a number of equally likely cases, it is usually a relatively simple matter to determine those probabilities that interest us, as will be seen in connection with the game of poker. This analysis fundamentally depends in every case on what has been called a "common-sense judgment." Fortunately, it is often possible to test the soundness of this judgment by appealing to experience. For common sense has a very questionable record in the history of science. It has been said that it might well be replaced by "uncommon sense," and this last has often, to the uninitiated, a striking resemblance to nonsense. However, in all of the problems we shall consider later on, the possible cases will impress us as so obviously similar that our skepticism is not likely to be aroused unduly.

There is a difficulty of a logical order in this definition of probability, due to the occurrence of the words *equally likely* as applied to the possible cases. Equally likely cases would seem to be those whose probabilities are equal, thus making the definition of probability depend on the idea of probability itself, a clear-cut example of a vicious circle. This is a little like defining a cow as a cowlike animal. It means that we must have some notion of what equally likely cases are before we tackle any concrete problem. In most of the situations to which we shall apply this definition of probability, however, our preliminary judgment as to the equally likely cases is based on mechanical considerations. In throwing a die it is the symmetrical construction of the die that leads us to believe that its sides are equally likely to turn up. If the die is "loaded," which means that it is not symmetrical, this is no longer true, as many craps players have found out, sometimes a little late. Or imagine a die of uniform material with six flat sides which are *unequal*. This mechanical change has knocked our definition completely out, for there are no equally likely cases into which the situation can be analyzed.

In drawing numbered balls from an urn we say that each of the possible draws of a single ball is as likely as any other, provided that the balls are identical in construction, and so well mixed that each draw may be thought of as entirely independent of preceding ones. Also in drawing one card from a deck of fifty-two cards we say that one card is as likely to appear as another, provided that the deck has been thoroughly shuffled, and the faces of the cards are hidden.

This instance brings to light a very important consideration:

a priori probabilities depend on the amount of knowledge assumed as given; they change when the assumed amount of pertinent knowledge is changed. If you are able to see some of the cards when making the draw, it is no longer possible to assume that all fifty-two cards are on the same footing. We cannot be sure that you are not influenced one way or another by what you see. Furthermore, the proviso that the deck be shuffled has no importance whatever unless you have some idea in advance of the order of the cards, as you would, for instance, if the deck were new.

Let us get back to games of chance and see how our rule for finding probabilities works in slightly more complicated situations. Suppose that instead of one die we throw two, as in the game commonly known as craps. What is the chance of throwing a total of 7? First of all we must know the total number of possible cases. In other words, we must count the number of distinct throws that can be made with two dice. Clearly there are thirty-six. For a given face of one die, say the 4, can be paired in six ways with the sides of the other die. There are six such pairings, each giving six distinct possible throws, so that the total number is thirty-six.

Before going any further, let us stop long enough to give each of our dice a coat of paint, one blue, the other red. For unless we have some way of distinguishing the two dice from each other, we are liable to fall into a serious error, as will be plain from what follows. We are now ready to list the *favorable cases,* in other words those throws that total 7. If we turn up the ace on the blue die and the six spot on the red die, the total is 7, and we have the first of the favorable cases. But we also obtain a total of 7 if we turn up the ace on the red die and the six spot on the blue die. Thus there are *two* favorable combinations involving the ace and the six spot.

This simple fact, so obvious when the dice are thought of as painted different colors, was once a real stumbling block to the theory of probability. If we continue the listing of the favorable cases, we can write the complete list in the form of a short table:

TABLE I

Blue	Red
1	6
6	1
2	5
5	2
3	4
4	3

Each line of this table indicates one favorable combination, so that there are six favorable cases. As the total possible cases number thirty-six, our fundamental rule tells us that the probability of throwing 7 with two dice is $\frac{6}{36}$, or $\frac{1}{6}$.

This result is correct, provided that each of the thirty-six possible cases is equally likely. To see that this is indeed the case, we remember that each of the six faces of a single die is as likely to turn up as another, and that throwing two dice *once* comes to the same thing as throwing one die *twice*, since the throws are independent of each other.

What is the probability of throwing 8 with two dice? The only change from the previous example is in the list of favorable cases. This time we get the table:

TABLE II

Blue	Red
2	6
6	2
3	5
5	3
4	4

There are five lines in the table, so that the probability of throwing 8 with two dice comes out $\frac{5}{36}$. Notice that the combination 4–4 can occur in only one way.

To find the probability of throwing any other total with two dice, all that is necessary is to list the favorable cases as above. The results in full are:

TABLE III

Total of Throw	Probability
2 or 12	$\frac{1}{36}$
3 or 11	$\frac{2}{36}$ (or $\frac{1}{18}$)
4 or 10	$\frac{3}{36}$ (or $\frac{1}{12}$)
5 or 9	$\frac{4}{36}$ (or $\frac{1}{9}$)
6 or 8	$\frac{5}{36}$
7	$\frac{6}{36}$ (or $\frac{1}{6}$)

These results can be checked by noting that the sum of the probabilities of all the possible cases should be equal to 1, since it is certain that one of them will occur. In the table each of the first five lines represents two cases, as indicated. If each of the probabilities in these lines is multiplied by 2, and the sum is taken, including the last line, the result is indeed $\frac{36}{36}$.

These figures have a direct application to the game of craps, and will be used in the discussion of that game in Chapter XIII.

It is sometimes easier to compute the probability that an event will *not* happen than the probability that it will. If the event is such that on each trial it either takes place or does not take place, the knowledge of either of these probabilities gives us the other at once. We merely subtract the known probability from 1. In order to use this rule it is necessary to exclude such possibilities as dead heats, drawn games, and so on. Its use is well illustrated by a simple problem from the two-dice game. We ask the chance of throwing either one or two aces in a double throw. The total possible cases are, of course, thirty-six. We must count the favorable cases. As every such case contains at least one ace, the unfavorable cases are those that contain no ace. The number of these latter is at once obtained by imagining a die with five equal sides, the ace being omitted, and counting the total possible cases in two throws with such a die. There are evidently five times five, or twenty-five cases, just as with a six-sided die there are six times six, or thirty-six cases. Thus the probability of *not* throwing an ace with a pair of dice is $^{25}/_{36}$. It follows that the probability of throwing either one or two aces is $1 - ^{25}/_{36}$, or $^{11}/_{36}$, since it is certain that one of two things will happen; either we throw one or more aces or we do not.

It is also easy to compute directly the probability of throwing either one or two aces with two dice. The favorable cases for the two dice are: ace and not an ace, not an ace and ace, ace and ace. We have again thought of the dice as though they were painted distinct colors, always listing one, say the blue, first. In this enumeration there are three favorable cases. In order to apply our definition of probability, however, we must have an analysis into *equally likely* cases, which the above are obviously not. The first contains five simple combinations of the sides of the two dice, the second five, and the third one, a total of eleven. The probability of throwing at least one ace therefore comes out $^{11}/_{36}$, in agreement with our previous result.

The greatest care in deciding which are the equally likely cases in a given problem is essential. In most cases we are able to attain what amounts to a moral certainty of the correctness of an analysis, once it is accurately made. This fact does not, however, prevent gross errors that result from too hastily assuming that the analysis is complete. Mistakes of this sort, and some others that are quite as elementary, seem to have a fatal attraction, not only for beginners in the subject, but for masters as well. In one of the best texts of recent years, for instance, it is stated that the probability of throwing at least one ace with two dice is $^1/_3$. We have just seen that the correct probability is $^{11}/_{36}$.

Among classic mistakes there is the one committed by

d'Alembert, the French mathematician and encyclopedist of the eighteenth century. For the following account of it I am indebted to Arne Fisher's well-known book *The Mathematical Theory of Probabilities:* Two players, A and B, are tossing a coin with the following rules: The coin is to be tossed twice. If heads appears on either toss, A wins; if heads does not appear, B wins. d'Alembert remarks that if the first toss is heads, then A has won and the game is over. On the other hand, if the first toss is tails it is necessary to toss again, and heads on this second toss means victory for A, while tails means victory for B. d'Alembert reasons that there are thus three cases, the first two of which are favorable to A, the third to B, so that the probability of A's winning is ⅔.

The error in this reasoning is the very one against which we have several times taken precautions. d'Alembert's three cases are not equally likely to occur. *The one in which heads appears on the first toss really represents two cases,* heads followed by heads, and heads followed by tails. Whether or not the game ends after the first toss is wholly beside the point in computing chances based upon equally likely cases.

To find the correct chances of A and B, let us begin by listing *all* the possible cases, indicating a toss of heads by H, a toss of tails by T. The complete list of equally likely cases is: HH, HT, TH, TT. In this list HH means that both the tosses are heads, and so on. Each of the first three cases contains at least one H, and therefore is favorable to A; only the last case is favorable to B. The correct probability that A will win is therefore ¾ or, in other words, the betting odds are 3 to 1 against B.

We have already referred to the fact that when we compute a probability the result depends on the amount of information that was given us in the first place. If two individuals have different amounts of knowledge of a situation and compute the probability of one of its component events, they may obtain quite different values, each of which is correct *relative to the particular individual.* A simple example will illustrate this point. In what is known as the "three-card game" the sharper (it usually *is* a sharper) places three cards face down on a table, announcing that exactly one of them is an ace. You are to guess which one it is. Assuming that the game is honest, you have exactly 1 chance in 3 of picking out the ace. If the deal was badly made, so that you saw one of the three cards enough to know that it is not the ace, your chance to win would become 1 in 2, while for those spectators who failed to notice the exposed card the chance would remain 1 in 3. But if the game is conducted by a cardsharp, another factor appears, the factor of deception or trickery, so that your chance of success is very small indeed. Any spectator aware of this fact will estimate the chances accordingly.

It is not to be thought that because probabilities are subjective in this sense, they have any the less significance or value. For if they are subjective in one sense, they are strictly objective in another. The probability of an event is the same for every individual who possesses the same amount of knowledge about it. Thus probability is in an entirely distinct category from such ideas as appreciation or artistic discrimination, which depend not only on the individual's knowledge, but in an essential way on the individual as a whole.

The knowledge that one has of a problem in probabilities is in fact a part of the statement of the problem. The more clarity and accuracy there are in the statement of the problem, the more these qualities may be expected to appear in the solution. In a game of pure chance complete information is presumably contained in the rules of play, so that, barring dishonesty, all players can be assumed to have the same knowledge, and the computed probabilities will apply equally to all.

CHAPTER VI

"*Heads or Tails*"

JUST about the simplest gambling game in the world is the one commonly called *heads or tails*. It is played by tossing a coin in the air; one face or the other must be uppermost when the coin comes to rest, and assuming that it is symmetrically made, the probability of either result must be the same. Clearly the probability is exactly ½ that it will be heads, and the same for tails.

This game sounds so simple that at first glance there would appear to be no reason for examining it further. But its very simplicity makes it one of the best illustrations of the fundamental principles of the theory of probability, and it contains, furthermore, a number of interesting problems which do not seem altogether elementary to those who come across them for the first time. We might ask, for instance: What is the probability that you will throw twenty heads before your opponent has thrown fifteen? We shall approach the game, however, from the standpoint of its relation to "red and black." In a roulette wheel manufactured for private play, as distinguished from professional play at a casino, half the compartments are colored red, the other half black, and perhaps the commonest form of play consists in betting on which color the marble will fall into next. With such a wheel there is no advantage for the "banker," and the game is therefore identical with "heads or tails."

Roulette, of course, is ordinarily played between the "bank" or capital of an establishment on the one hand, and the individual gambler or gamblers on the other. The bank undertakes to accept all bets from all comers, but it limits the size of the bets in both directions. That is, you may not bet less than a certain minimum amount nor more than a certain maximum. As we shall see, this restriction has important consequences for the players, as well as for the bank.

Before examining the probabilities involved in "red and black," it is well to explain that in speaking of it here, and in more detail in Chapter XII, we shall be talking about roulette as it is played at Monte Carlo, where the wheels have only one "0" compartment. The presence of this 0 compartment on the wheel means, naturally, that neither red nor black is quite an even chance—the percentage in favor of the house is actually 1.35, as computed in the light of the special Monte Carlo rules governing the 0. The details of the calculation will be given in Chapter XII. In the long run, then, the players as a group lose just 1.35 per cent of the total amounts they stake.

This is a relatively low percentage compared with that at many American casinos, where the wheels have both a 0 and a 00 compartment, and both represent outright losses to the players. For that reason, the play here discussed will be based on the Monte Carlo wheels, but the conclusions arrived at can easily be modified to apply to other forms of the game.

It is hardly necessary to add that what is said about "red and black" at roulette will apply with equal force to the other two forms of approximately "even" bets at roulette—in which the players gamble that the marble will stop in an even-numbered or an odd-numbered compartment, or that it will stop in a compartment with a number from 1 to 18, inclusive, or a number from 19 to 36 (there being, besides the 0, a total of thirty-six compartments on a Monte Carlo wheel). In the case of these other types of "even" betting the house or bank percentage, 1.35, is naturally the same as in the case of "red and black."

Although this small advantage of the bank is decisive in the long run, we shall leave it out of account for the present and assume that there is no distinction between the "even" chances of roulette (assuming that the roulette wheel is perfectly constructed and honestly operated) and the game of heads or tails, so that in discussing the one we are also discussing the other. In Chapter XII, account will be taken of the fact that the odds of the game favor the bank.

Most players who place their money on the even chances in roulette do so because they believe that by following some particular method of play, or "system," they are likely to emerge a winner, although the amount of the win may be comparatively small. On the other hand, in betting on the longer chances, where systematic play is in the nature of things almost impossible, extraordinary runs of luck have their full sweep, in both directions. Inconspicuous behind the glare of dazzling individual exploits, the slow workings of the laws of chance are apt to be overlooked. The player tends to attribute his good or bad fortune to the state of his relations with the goddess of luck, relations subject to very sudden and seemingly capricious fluctuations.

A large number of "systems" of play for the even chances have been invented. All of them depend on combinations of a small number of ideas. On the desk as I write is a list of over one hundred such systems (among which is apt to be the one "invented" last year by your friend, who considers it too valuable to be disclosed) .

Systems can be divided into two classes: those that have some effect on the probability of gain or loss *during short periods of play,* and those that have no effect whatever.

There are no systems that have any effect in the long run.

It is impossible (or nearly so) to learn this fact from the experience of one individual, and as very few persons can or will make a large-scale study of the matter, these systems, when demolished, spring up again like the heads of Hydra before the assault of Hercules. It is a curious fact that the partisan of one system usually has a profound mistrust of all others; only the system that he is using at the moment has merit and soundness—all others are thoroughly exploded. So we shall attempt to indicate the principles on which such systems rest, without hoping to convince the system player that the value of his particular system is illusory.

One of the oldest and best known roulette systems is the one called the Martingale. In principle it closely resembles the betting practice commonly known as "double or nothing." If you win $5 from a friend, and wish to be particularly friendly about it, you may say: "Double or nothing on this toss of the coin." If he wins, the $5 debt is canceled. If you win, the situation is twice as bad as before, from the friendly point of view; but you can go on with the double-or-nothing idea until the account is squared.

The rule for playing a Martingale against the bank is as follows: You begin by placing a sum of money, say $1, on either heads or tails, red or black. If you win the toss, you again place $1 on either heads or tails, as fancy dictates. But if you lose the first toss, your next bet is for $2, and if you lose both the first and second tosses, your third bet is for $4 and so on. In other words, whenever you lose you double your preceding bet. Whenever you win, your next bet is invariably $1. If you have sufficient capital to keep on doubling after each unfavorable toss, and if you win a toss before the maximum permissible stake is reached, it is not hard to see that you will have won $1 for each toss of the series that was favorable. Suppose, for example, that you bet always on heads, and that the series consists of one hundred tosses, fifty-three of which (including the last toss) were heads. Then your net result is a win of $53.

If the player were assured, before he sat down to play, that he would never lose enough times in succession to be forced to discontinue doubling because he has reached the maximum stake, the Martingale would be a very lucrative game. Such an assurance, however, would be flatly contradicted by the laws of chance. In the long run, as we shall see later, the risk that the player takes exactly balances his prospects for profits. In roulette, where the odds of the game favor the bank, this risk is *not* balanced.

At this point it is convenient to introduce two terms that describe important characteristics of roulette. If we divide the minimum stake by the maximum permissible stake, we obtain a number that is of particular importance in system play. We shall call it the "pitch" of the game, for it measures the *tension* of system play. Similarly, the minimum stake is a measure of the *steepness* of the game and might well be called the "scale."

At Monte Carlo, in the outer rooms, the maximum stake is 12,000 francs, the minimum 10 francs.* The "pitch" is therefore 1/1,200; the "scale" is 10 francs. If a player makes an initial bet of 10 francs and doubles the bet after each successive loss, the series of bets runs: 10, 20, 40, 80, 160, 320, 640, 1,280, 2,560, 5,120, 10,240. Thus he can double his initial bet ten times before reaching the maximum. If he loses on each occasion his total loss has come to 20,470 francs (found by doubling the amount of his last bet and subtracting the initial bet, which is 10 francs). If the minimum stake was 1 franc, instead of 10 francs, and if the player started with an initial bet of 1 franc, he could double three additional times (in other words, thirteen times), the last bet being 8,192 francs, and the total loss 16,383 francs. This indicates the importance of the amount of the minimum stake, and the reason why system players start out with the smallest permissible bet. The extent of their possible winnings is reduced in this way, but the chance of complete disaster in a fixed number of plays is also greatly reduced. It is the usual custom in describing "systems" of play, and a very convenient one, to take the initial bet (usually equal to the minimum stake) as the unit and to express all bets as a

* This was so, at least, prior to World War II. If there has been a change, the figures in these pages are easily corrected.

certain number of units. Thus in Monte Carlo roulette the unit is ordinarily 10 francs; a bet of 100 units means a bet of 1,000 francs.

Suppose that we imitate one of the even chances of roulette as closely as possible by tossing a coin one time after another. Each toss is *independent* of all preceding tosses. This means precisely this: the probability of heads (or of tails) is always ½, regardless of how many or how few heads have gone before. This is the point that was discussed at some length in Chapter IV, where the fallacies underlying the doctrine of "the maturity of the chances" were brought out. An understanding of it is absolutely essential in connection with roulette, if the grossest sorts of errors are to be avoided. If a player believes in the "maturity of the chances," he is logically justified in expecting to win on a system that calls for increased bets after successive losses. "For," he must argue, "after I have lost several times in succession, the chance of my losing again is *less* than 1 in 2; but the house is offering to bet at the same odds (very nearly even) as when I win several times in succession. Now I place *large* bets when I have lost, and *small* bets when I have won. Therefore the odds are decidedly in my favor."

If the simple fact, so frequently proved by experience, that the chance of heads is always 1 in 2, no matter whether there have been ten, or twenty, or any other number of consecutive heads, were once grasped, together with its logical consequences, the majority of the roulette systems would "softly and suddenly vanish away," for their popularity depends on the hope of their producing a profit in the long run.

On the first toss of a game of heads or tails there are two equally likely results, heads and tails, and the probability of each is ½. Let us represent a toss of heads by H, one of tails by T, and consider a series of two tosses. The equally likely cases are HH, HT, TH, TT, and the probability of each is ¼. For three tosses the cases are HHH, HHT, HTH, HTT,

THH, THT, TTH, TTT, eight in all, so that the probability of each is $\frac{1}{8}$.

We obtained this list of cases for three tosses from the corresponding one for two tosses by adding successively an H and a T to each of the earlier cases, thus doubling their number. This rule is perfectly general. To get the list of possible cases for four tosses from that for three we add in turn H and T to the end of each, giving sixteen cases in all. For example, the first case listed for three tosses is HHH, and by adding first an H and then a T we get HHHH and HHHT. The probability of each is $\frac{1}{16}$. In any number of tosses, say n, the probability of any particular throw, such as n heads in succession, is a fraction whose numerator is 1, and whose denominator is 2 multiplied by itself n times. This is written $\frac{1}{2}^n$.

We have thus found the probability of any particular result of a series of throws of any length, using directly the definition of probability as the number of favorable cases divided by the total number of cases, all being equally likely. This manner of procedure has one disadvantage; in certain more complicated problems it becomes very cumbersome, sometimes involving an extravagant amount of labor. It is often possible to break the calculation into a series of easy steps. Tossing a coin will serve very well to illustrate this method: Suppose we wish to find the probability of tossing three consecutive heads in three tosses. The probability of heads on the first toss is $\frac{1}{2}$. The second toss is entirely independent of the first, so that *after the first toss is made,* the probability of heads on the second remains equal to $\frac{1}{2}$. The probability that these two events (heads on each toss) both take place is $\frac{1}{2}$ times $\frac{1}{2}$, or $\frac{1}{4}$. One more repetition of the same argument shows that the probability of three heads in three tosses is $\frac{1}{8}$. This agrees with the corresponding result in the previous paragraph, where all of the possible cases were listed.

This way of finding a probability one step at a time is often of value and can be expressed in the form of the following rule:

The probability that two independent events will both take place can be found by multiplying together the separate probabilities of the events.

Independent events are events that have no influence on each other. We have just given illustrations of such events and of the use of the above rule in connection with repeated tosses of a coin. The same is true of repeated throws of a die. The chance of throwing two sixes in two throws of a die is $\frac{1}{6}$ times $\frac{1}{6}$, or $\frac{1}{36}$. This must agree with the chance of throwing a total of 12 on a single throw of two dice, for which we found the same value in Table III on page 50.

When two events are not independent, the use of the above rule would lead to serious errors. Suppose that four playing cards are laid face down before you, and you know that exactly two are aces. Your chance of getting an ace on the first draw is $\frac{2}{4}$, or $\frac{1}{2}$; for there are two favorable cases out of a total of four cases (all equally likely). Now what is your chance of drawing both the aces on two draws? If the drawings were independent of each other, the above rule would say that it is $\frac{1}{2}$ times $\frac{1}{2}$, or $\frac{1}{4}$. But they are *not* independent. Suppose that you have already drawn an ace on your first effort. There remain three cards, *one* of which is an ace, so that your chance on the second draw is 1 in 3, not 1 in 2. The chance that you will pick two aces in as many draws is therefore $\frac{1}{2}$ times $\frac{1}{3}$, or $\frac{1}{6}$, instead of $\frac{1}{4}$. In other words, the probability of the second event must be found on the assumption that the *first event has already taken place.* The general rule in such cases is:

The probability that two successive events both take place is equal to the probability of the first event multiplied by that of the second, the latter being computed on the assumption that the first event has already taken place.

It should be noticed that this second rule does not mention the independence or dependence of the events. The reason is that it works for either variety. If the events are independent, the probability of the second is not changed by the occurrence of the first, and this rule becomes the same as the previous one.

While speaking of the various sorts of relations that there can be between a pair of events, we should complete the story by mentioning the third alternative. Sometimes the occurrence of one event prevents the possibility of the other, and vice versa. They are then called *mutually exclusive* events. The simplest example is heads and tails, the coin being tossed once. If one occurs, the other cannot possibly occur. On the other hand, one or the other is bound to happen, and the probability of anything that is certain is 1, as we have seen. But this is what we get if we add the probability of heads, which is $\frac{1}{2}$, to the probability of tails, which is also $\frac{1}{2}$. This is always true of mutually exclusive events. The probability that one or another of several such events will happen is found by adding the separate probabilities of the events. For example, the probability of throwing either a 6 or a 5 with one die is $\frac{1}{6}$ plus $\frac{1}{6}$, or $\frac{1}{3}$. There is a similar situation in any game of chance in which one and only one person wins. The probability that someone of a certain group of players will win is simply the sum of the separate probabilities of the players in question.

It is now time to get back to the game of heads or tails in which one player is playing against a "bank." Having the right to place his bets in any way that pleases him, he has decided to play according to the Martingale system. He begins by placing $1 on either heads or tails, let us say heads, doubling his bet after each loss, and always betting $1 after a win. To avoid mistakes the player keeps a score sheet, entering the number of dollars bet, the result of the toss, the number of dollars won or lost on each toss, and a running total, in which he uses a minus sign to indicate that he is out of pocket. Let us assume that he is wise enough to realize that his prospects are not altered one way or the other by shifting his affections from heads to tails and back, either regularly or irregularly, and so eliminates an

TABLE IV

Player Wins on Heads (H), Loses on Tails (T)

Result of Toss	Bet	Win	Lose	Total ($)	Result of Toss	Bet	Win	Lose	Total ($)
T	1		1	−1	T	16		16	−18
T	2		2	−3	H	32	32		14
T	4		4	−7	T	1		1	13
H	8	8		1	H	2	2		15
T	1		1	—	T	1		1	14
T	2		2	−2	T	2		2	12
T	4		4	−6	H	4	4		16
H	8	8		2	H	1	1		17
T	1		1	1	H	1	1		18
H	2	2		3	T	1		1	17
T	1		1	2	T	2		2	15
H	2	2		4	T	4		4	11
T	1		1	3	H	8	8		19
T	2		2	1	H	1	1		20
H	4	4		5	T	1		1	19
T	1		1	4	T	2		2	17
H	2	2		6	T	4		4	13
T	1		1	5	T	8		8	5
H	2	2		7	H	16	16		21
T	1		1	6	T	1		1	20
T	2		2	4	H	2	2		22
T	4		4	—	H	1	1		23
H	8	8		8	H	1	1		24
T	1		1	7	T	1		1	23
T	2		2	5	H	2	2		25
T	4		4	1	H	1	1		26
T	8		8	−7	T	1		1	25
T	16		16	−23	H	2	2		27
H	32	32		9	H	1	1		28
H	1	1		10	H	1	1		29
H	1	1		11	H	1	1		30
T	1		1	10	H	1	1		31
H	2	2		12	H	1	1		32
T	1		1	11	H	1	1		33
T	2		2	9	T	1		1	32
H	4	4		13	H	2	2		34
T	1		1	12	T	1		1	33
T	2		2	10	H	2	2		35
T	4		4	6	H	1	1		36
T	8		8	−2	H	1	1		37

extra column on his sheet by betting always on heads. He loses, then, only on long runs of tails on which either his purse is exhausted, or the maximum, representing the limit of the bank's patience, is reached. Let us glance over his score card on page 63 (based on an actual series of eighty tosses), which has a happy ending, for neither of these disasters overtakes him.

The player has gained $1 each time that heads appeared, or thirty-seven times. His highest bet was $32 after five consecutive losses. If he had been backing tails instead of heads, he would have encountered one adverse run of 7, requiring a bet of $128 on the next succeeding toss, making a total of $255 laid out during the run. To find the extent of the demand made upon his playing capital, which gamblers call the "strain," his profit at the beginning of the adverse run must be subtracted from this figure. In this case it is $41, so that the "strain" would be $214.

If the bank placed no limit on bets, and if the player's resources were sufficient to meet the strain encountered during the play, he would win one unit each time he won a toss, and as in the long run he can expect to win half the time, this comes to $\frac{1}{2}$ unit per toss. But the *ifs* are all-important, enough so, it will appear, just to balance the ledger. For players of equal resources the game of heads or tails presents equal prospects of winning or losing, and no amount of shifting about of bets can alter this fact. In playing a system of the type of the Martingale what the player in fact does is this: *In a limited series of tosses his chance of a small gain is large, that of a large loss small.* The two balance each other.

We have seen that the chance of throwing two successive heads in two tosses is $\frac{1}{4}$, and that in general the chance of throwing any number n of successive heads in n tosses is $\frac{1}{2}^n$. What we need to know next is the chance of throwing a number of heads in succession during a long series of tosses. When the toss of a head is preceded and followed by tails we shall call it an *intermittence* on heads, following a convenient terminology of roulette. There is an intermittence on heads whenever the three successive tosses THT are encountered, one on tails in case it is HTH. In Table IV there are eleven intermittences on tails, fourteen on heads. By a *run* of heads we mean a number of consecutive heads preceded and followed by tails. For instance, a run of 4 on heads is given by THHHHT; it does not matter how many other tosses went before or came after, or what the other tosses are.

In a long series of tosses how many intermittences on heads ought we to expect? Take any toss of the series except the first or the last (for these the word *intermittence* has no meaning), say the second. In order for this to be an intermittence three things must be true: Toss number 2 must be heads, toss number 1 tails, and toss number 3 tails. Since the tosses are independent of each other and the probability of each of these events is $\frac{1}{2}$, the probability that all three take place is $\frac{1}{2}$ times $\frac{1}{2}$ times $\frac{1}{2}$, or $\frac{1}{8}$. If we had considered a run of two heads instead, that is THHT, we should have had an extra factor of $\frac{1}{2}$, so that the result would be $\frac{1}{16}$. For a run of any number n of consecutive heads the probability is $\frac{1}{2}^{n+2}$. To illustrate these results consider a series of 4,096 tosses. Noting that a run of 1 and an intermittence are the same thing, the number of tosses on which we should expect the indicated run is given by the following table:

TABLE V

EXPECTED NUMBER OF RUNS ON HEADS (OR TAILS)

Run	1	2	3	4	5	6	7	8	9	10	more than 10
No.	512	256	128	64	32	16	8	4	2	1	1

Number of tosses 4,096.

Notice that according to the table the number of runs of more than ten heads is the same as the number of runs of exactly ten heads. This is true of the table no matter where we cut off. For example, the table indicates that we should expect 8+4+2+1+1 or 16 runs of more than 6, and this is exactly the expected number of runs of 6.

In discussing intermittences and runs, the first and last tosses of a series were carefully excluded, as the reader has noticed. For a long series of tosses this has no appreciable effect, but for short series like that in Table IV it may require the omission of several tosses or an appreciable fraction of the whole. The series in Table IV begins, in fact, with three tails. In listing the runs how should these cases be included? It is worth while to answer this question because it illustrates a rather general statistical practice.

It would evidently be wrong to classify it as a run of 3 on tails, since doing so would amount to assuming a previous toss of heads. Now imagine that there *was* a previous toss. The probability that it was heads is ½. In a sense therefore we have *one half* of a run of 3, so that we shall add ½ to the number of runs of 3. If the previous toss was tails the run remains "open" at one end, and it is necessary to imagine a second toss preceding the first one. The chance that this last was heads, and that the first was tails, thus completing a run of 4, is ½ times ½ or ¼. Therefore we credit ¼ to the number of runs of 4. Continuing, we credit ⅛ to the runs of 5, 1/16 to the runs of 6, and so on. If the number of credits is added up it is ½+¼+⅛+ and so on, which is equal to 1. We have therefore added 1 to the number of runs, which is just as it should be. It turns out that we have also added 1 to the total number of tosses, but that is of small importance.

Let us now list the runs in Table IV and compare their number with the normal, as found above. Although a series of eighty tosses is extremely short, and only a very rough agreement between theory and practice should be expected, it will be seen that even in so unfavorable a case the correspondence is striking. The reader who is interested in pursuing the matter further can easily construct a table for himself, its length depending only on his ardor in coin tossing. Here is the listing:

TABLE VI

Run	1	2	3	4	5	6	7	more than 7
Table IV	25	5	7	$1\frac{1}{2}$	$2\frac{1}{4}$	$\frac{1}{8}$	$1\frac{7}{16}$	$\frac{1}{16}$
Theory	20	10	5	$2\frac{1}{2}$	$1\frac{1}{4}$	$\frac{5}{8}$	$\frac{5}{16}$	$\frac{5}{16}$

Number of tosses 80.

This table shows the number of runs of the length indicated in the top line, in the eighty tosses that make up Table IV, as compared with the number that theory indicates should occur in the long run, based on Table V. Since we are listing runs on both heads and tails, the numbers given in Table V must be doubled, as well as adjusted for a total of eighty tosses. Table VI tells us, for example, that the eighty tosses contained twenty-five runs of one (intermittences), while theory predicts twenty. The fractional values in the second line come from the first and the last series of tosses in Table IV, which we have "spread" according to the principles just explained. In this instance the series starts with three tails, so that we must make the entry ½ in the column headed "*3*," and so on. This same column gets a credit of ½ from the final three tosses of heads, so that the number of runs of 3 is increased by one.

In the game illustrated in Table IV the player added $1 to his gains each time that heads appeared, and could go on in this monotonous but lucrative fashion as long as his capital was sufficient to carry him through the adverse runs, always assuming that the bank places no maximum limit on bets. Disregarding this last, there is always a practical limit to the play, since the player's resources, no matter how great, are necessarily limited, and *if he continues long enough* he is certain to encounter an adverse run long enough to bankrupt him. Of course disaster may be slow in arriving. It all depends on his rate of play. In the later chapter on roulette we shall find out just what the chance of disaster is for various periods of play.

Why, then, does the bank find it necessary to put a limit on the size of the bets, and a relatively small one at that? The Baron de Rothschild is said to have once remarked to M. Blanc, founder of the Monte Carlo casino, "Take off your limit and I will play with you as long as you like." Why was this rather rash challenge not accepted? Of course it is understood that in practice the bank has always a small advantage in the chances of the game, and that it is from this source that the profits are derived. These profits are extremely probable, and when play is heavy the percentage of profit to the total amount staked is fairly constant over long intervals of time. But like all other products of chance it is subject to sudden fluctuations, sometimes of considerable magnitude, as when some fortunate player succeeds in "breaking the bank." The opposite case, when the bank is fortunate enough to break (and in this case the word may mean what is says) a large number of players in a short interval, is not so widely advertised. Now the effect of limiting individual bets is to decrease the extent of these inevitable fluctuations in both directions, and this decreases the slight risk that is always present that an unprecedented run of luck in favor of the players will draw too heavily on the bank's cash reserves, or even ruin it. Suppose, for instance, that not one but fifty Barons de Rothschild pooled their resources in a combined attack on the bank. For it must be remembered that the bank has set itself up against the whole public, and while it may prefer to engage them in single combat, it could not conceivably prevent the forming of syndicates. In fact, at Monte Carlo a good deal of the heavier play is admittedly with capital raised by private syndicates. On the other hand, the effect of the limit on individual bets is to decrease the total amount of the stakes, and therefore of the bank's profits, which are proportional, in the long run, to the total stakes. Clearly, the bank is not inclined to make this limit lower than necessary.

It will be seen in Chapter VIII that even if a syndicate with very large resources were organized, and if the bank were to remove its maximum limit on bets, the chance of the bank's ultimate ruin would still be small. Even so, it would be enormous compared to what it is with the present limit.

What if such a syndicate were to adopt one of the various roulette systems, the Martingale, for instance? Would it not, in the absence of a limit, be able to win considerable sums with almost no risk of loss? The answer is that if its capital were enormous, the risk of losing it would indeed be very small during, say, a few years of play. But the amount of gain would also be small compared to the amount of capital, so that, looked at from a business point of view, the return on the investment would be unattractive in the extreme. On the other hand, the prospect would also be unattractive from the bank's point of view. For although the bank could expect to win in the end, the "end" might be excessively slow in coming, and in the meantime its losses would mount up. This is an example of a risk that is *unfavorable to both parties.* If someone offered you 5 to 1 odds against your throwing heads, your stake being the sum of your worldly possessions, you would be as foolish in accepting the bet as he was in offering it.

Let us leave out of account the fact that at roulette the "even" chances are not quite even, and consider the outlook for the syndicate a little more closely. So far as I have been able to discover, the longest run on record is one of twenty-eight even numbers, which happened at Monte Carlo. If our syndicate had been playing a Martingale at this table at the time, and if it had been playing "odd and even" (rather than one of the other two even-chance games) , there is still no more reason to assume that it would be betting on odd than on even, and only in the former case would this run have been unfavorable.

How often should we expect a run of 28 *or more* to occur according to the laws of chance? We have already computed that on the average a run of 28 or more heads (or tails) will appear once in 2^{29} tosses. If runs on both heads and tails are taken account of, the expected frequency becomes twice in 2^{29} tosses, or once in 2^{28} tosses. We therefore expect a run of 28 or more odd or even numbers at roulette once in 2^{28} whirls of the wheel, or *coups,* to use the convenient French term; that is to say, once in 268,435,456 *coups.*

At each table there are three different even chances being played at the same time, and these chances are completely independent of each other, so that the number of plays is in reality multiplied by 3. We therefore expect a run of 28 or more at a table three times as frequently as when only one of these three sets of chances is taken account of, or once in 89,-478,485 *coups.*

A roulette table at Monte Carlo averages, I believe, about 480 *coups* per day, or about 168,000 per year, assuming play to go on something like 350 days of the year. Thus it takes roughly six years to play 1,000,000 *coups* at a table. We should therefore expect a table to produce a run of 28 or more on some one of the various even chances once in 534 years. It is difficult to estimate the number of tables in operation over a period of years; perhaps four is a reasonable estimate. If so, a run of 28 or more is to be expected on one or another of the four tables once in about 134 years. If five tables is a better estimate, this becomes once in 107 years. The fact that there was one run of 28 on one of the even chances at Monte Carlo during the first sixty-eight or sixty-nine years of its existence is therefore not remarkable.

Suppose that the syndicate played a Martingale at one table on one set of even chances, say red and black, and that it had sufficient capital to withstand an adverse run of 28, but not one of 29. According to the calculation just made, a run of 29 or more on one of the chances, say black, is to be expected about once in 6,442 years. That is the interval of time that has elapsed since the first Pharaoh of Egypt ruled, and the members of the syndicate might well feel that their risk in a few years of play would be very small. When the amount of this risk is taken account of, nevertheless, it appears that their expectation of loss balances that of profit, a statement that will be given a precise meaning later. Also it is to be remembered that we have neglected the small extra chance in favor of the bank, which makes matters worse for them.

In order to withstand an adverse run of 28, the syndicate would have to have a capital of $2^{29}-1$ units of 10 francs each, since after the twenty-eighth consecutive loss it would be necessary to double the preceding bet. This comes to the neat sum of 5,368,709,110 francs. Now the profit in a Martingale is, as we have seen, about ½ unit per *coup* played, which would come to 83,500 units per year, or an annual return of 835,000 francs. Figured against the above amount of capital this gives a yearly profit of .016 per cent, hardly an attractive rate of return on the investment.

The syndicate could, however, do very much better than this by taking advantage of the fundamental principle of the game, namely, that under all conditions the chance of red (or of black) is 1 in 2. It would not be necessary to keep the whole of this large sum in cash; the greater part of it could be invested in any form easily convertible into cash. Suppose, for example, that enough currency were kept on hand to withstand an adverse run of 20, but not more; this requires but 20,971,510 francs, and an adverse run of 21 or more will occur on the average once in about six years. Figured against this amount the profit is 4 per cent. Now if an adverse run of more than 20 is met with, it is necessary merely to interrupt play long enough to obtain sufficient cash to carry on.

This last statement impresses almost everyone who is not thoroughly at home in the theory of chance as sheer nonsense. Even the reader who has followed our previous discussions of related questions may have a vague, uneasy feeling that something is wrong. It is the ghost of the "maturity of the chances" doctrine, and we must make one last valiant effort to lay it for all time: You are representing the syndicate, let us say, and are betting always on red, doubling each time. There have been twenty consecutive blacks. Now the chance of red on the next toss (we return temporarily to the language of heads or tails) is either 1 in 2 or it is *not* 1 in 2. If it *is* 1 in 2, it cannot make the slightest difference whether you bet on the next toss or skip any number whatever of tosses, for red has the same chance on each. If you maintain that the chance is *not* 1 in 2, it is because you believe that the coin—or wheel—has contracted a debt to the backers of red, which it remembers and is about to pay off. In other words, you are casting a vote for that completely erroneous doctrine our old friend (or enemy), the "maturity of the chances." You will have to admit that interrupting the play has absolutely no effect on your prospects, one way or the other.

If the bank placed no maximum limit on the bets, it would be subject to another kind of attack based on a system that is the opposite of the one we have been discussing, and called for that reason the anti-Martingale. In this system the player doubles his bet after each *win* and returns to a bet of one unit after each *loss*. Before beginning the play he has made up his mind that if he wins a certain number of times in succession, say ten, he will withdraw his winnings from the play and his next bet will be for one unit. It is clear that his situation is the reverse of that of a player who is conducting a Martingale. The latter wins ½ unit per *coup* unless he encounters a long adverse run; the anti-Martingale player loses ½ unit per *coup* unless he encounters a long favorable run. In a short period at the table he has a *small* chance of a relatively large win and a *large* chance of a relatively small loss. We shall see in Chapter XII how these methods of play can be modified in almost innumerable ways, giving rise to a correspondingly large number of roulette systems.

The point we wish to make here is that in the anti-Martingale system the player's wins are entirely financed by the bank. For he merely leaves his original bet of one unit to double when he wins, each successive bet being made with the money previously won from the bank. With no limit an ambitious player could decide to leave his stakes until they had doubled twenty times, for instance (in the language of roulette this is called a *paroli* of 20), whereupon he would withdraw from the game no less than 20,971,510 francs. Now a run of 20 on some chance at some one of the tables is to be expected about once every six months, and as any number of players could be conducting this type of game at the same time, the bank, under these conditions, could not feel entirely secure. And security is quite as important to the bank, in a business sense, as it is to an insurance company. Hence, the bank's necessity for a maximum limit.

CHAPTER VII

Betting and Expectation

To SOME people it may seem undignified, or even pernicious, to discuss the theory of probability in terms of gambling and betting, as we have been doing at the beginning of this book. Because gambling has brought suffering and ruin to many a family through the ages, a reputable body of opinion has grown up that it is an evil practice, in short, that it is immoral. Of course we are not concerned here with any other aspect of the practice of gambling than the framework which it provides for certain principles of value in the theory of probability. Even those who disapprove of the practice probably know enough about gambling to understand illustrations drawn from it.

But though we are not here concerned one way or the other with ethics, it is interesting to remark that the sort of uncertainty in this world which makes betting a possibility appears to be fundamental to life as we know it. Remove that kind of uncertainty and you will have a world utterly unlike the one in which we now live. Such a world would be of extreme simplicity, regulated by laws thoroughly comprehensible to its inhabitants. These laws would, since chance would be no part of their structure, give a definite, accessible answer to questions about the future; the inhabitants of this non-chance world would therefore know the future even more precisely than we know our own pasts.

The most bizarre consequences, from our point of view, would stem from this condition of life. Ideas like probability would have no place at all. Betting on future events would be like betting on the game of checkers that you have just finished. There would be no such thing as an accident, and anyone who had the illusion of free will would be classified as a moron. A person would see very clearly that on the following day he was certain to meet with an unfavorable event (we would call it an accident), but could do nothing to avoid it, any efforts in that direction having been equally well foreseen and included in the calculations of the future.

Whether our own complex world has in principle close analogies to the simple one we have just conjured up, we do not know. At least we are protected from the distressing condition of its inhabitants by mountains of complexities that give little sign of crumbling away as a result of our analytical borings and tunnelings. And then there is the very definite possibility already referred to that the laws of our world will turn out to be statistical in character. In that case future events will always be coupled with probabilities; the element of chance will always be present.

On the other hand, if science could be imagined in some distant future to be on the point of achieving its old ambition of embracing all events in a scheme of nonstatistical or determinate laws, men would be forced to legislate hurriedly against its further practice, while they still retained enough of the illusion of free will to be able to do so! We need feel no alarm over the prospect. Even if such a set of laws could conceivably be discovered, there would have to be a miraculous change in man's brain before he could do much with them; for the number of events to be considered is so large that we can form no idea of it.

So, if betting is an evil by-product of the world in which we actually do happen to live, we may at least become more reconciled to the fact that our world is of a sort that permits the abuse to be practiced.

In relation to the theory of chances, betting is something like the advanced guard of an army. It can poke into all sorts of odd places and take chances that would be out of the question for the main body of troops. When two betters settle the terms of a wager, they are in effect settling a problem having to do with chances and, though their solution may be entirely incorrect, it is never ambiguous. Which is saying a good deal for it. When these betters agree upon odds of, say, 7 to 2, there can be no doubt concerning their joint opinion of the chances involved. Chances, be it noted, which relate to something in the future, in the past, or to a matter of mere opinion, but always with the indispensable condition that the matter in dispute shall admit of verification.

When two individuals make a bet, each is saying in substance that he considers the odds at least not unfavorable to his cause. (If there are altruistic people who make bets when they consider the odds unfavorable, we shall have to leave them out of account.) In the majority of actual wagers it is impossible to find out, either before or after the event in question, whether the odds were fair, and in some cases this question may even have no clear meaning. If A offers B 2-to-1 odds that he cannot make the next golf hole in four strokes, for instance, his offer may be based on an observation of B's game that applies only at the time of the wager. It does not at all follow that he will always offer the same odds. In such a case, although the wager itself is easily settled, there is no way to determine whether the odds were "fair" or not. On the other hand, whenever the bet concerns an event that admits of repetition under the same conditions, or is one for which the probability can be computed in advance, the fair betting odds can be determined. In many games of chance both these methods are applicable.

In introducing the idea of chance or probability, we took it for granted that everyone is familiar with the simple relation between the odds for or against an event and the chance that the event will happen. If the odds are even, or 1 to 1, the chance of the event is 1 in 2; if the odds are 2 to 1 against the event, its chance is 1 in 3, and so on. We shall merely illustrate the general rule by considering an event whose probability is $5/12$. Then the probability that the event will *not* occur is $7/12$. The two numerators, then, indicate the betting odds, which in this case are 7 to 5 against the event.

We have seen in Chapter V that if the probability of an event is $5/12$, the "law of large numbers" tells us that of a long series of repetitions or trials, about $5/12$ will give the event in question. Applied to betting this means that when the odds are fixed according to the above rule, the two betters will come out about even in the long run, if the bet is repeated. (The precise meaning of this statement, given in Chapter V, should be kept constantly in mind.) This is one way of saying that the odds are fair.

If A gives B odds of 5 to 2 on an event with probability $2/7$, the odds are fair, and A and B will come out about even in the long run. It comes to the same thing to say that they will come out exactly even in a series of *seven* repetitions, on two of which the event in question occurs. If A has put up $5 against B's $2, in such a series A wins five times and loses twice, so that he comes out exactly even.

There is another way to express the fairness or unfairness of odds which introduces a new idea that is of value in any situation involving a future monetary risk, whether it is a bet, a game of chance, or a business transaction. This new idea is that of *mathematical expectation*. Actually, the adjective is superfluous; so we shall call it simply the *expectation*. If you have 1 chance in 10 of winning $50, your expectation is $5. If you were to sell your risk to another person, that is the value that you would place on it, provided, of course, that the monetary consideration is the only one present.

The rule for finding your expectation is always the same as in this simple example: *Multiply the probability that the event will be favorable by the amount of money that you will take in if it is.*

We can now express what we mean by fair betting odds as

follows: Multiply the probability of your winning the bet by the total amount of the stakes. If the result is equal to the stake that you put up, the odds are fair to both parties; if the result is less than this amount, the odds are unfair to you; if more, unfair to your opponent. In the example given above, in which A gives B odds of 5 to 2, A's expectation is found by multiplying his chance to win, which is $5/7$, by the total amount of the stakes, which is $7, so that it is $5. As this is the amount that A contributed to the stakes, the bet is fair.

If A has 1 chance in 5 of winning $100, his expectation is $20. Using once again the "law of large numbers," we may also interpret this as meaning that A will expect an average win of about $20 on each bet, provided the event in question is repeated a very large number of times. He can afford to pay $20 at each repetition for the privilege of taking this risk; if he pays any less amount he can look forward to a profit.

We can say, then, that a bet, or a gambling game, or, more generally, a risk of any sort, is fair to all parties to it if each one risks a sum exactly equal to his individual expectation.

So far we have spoken only of profits. What about losses? If you have 1 chance in 20 of losing $100, we shall say that your *expectation of loss* is $5, or we shall say that your expectation is minus $5—an accountant would use red ink for it. Whenever the opposite is not explicitly indicated, however, we shall use the word *expectation* as referring to a profit.

The paramount fact that makes it valuable to use the idea of expectation in connection with risk is this: If you are faced with several possibilities of gain (or of loss), your total expectation is found simply by adding together the various expectations corresponding to the various possibilities, each computed as though the others did not exist. It makes no difference whether the events involved are independent, dependent, or mutually exclusive. We have seen, on the contrary, that the *probabilities* of events may be added only when the events are mutually exclusive; in other cases absurdities would result, such, for instance, as probabilities greater than one.

I agree to pay Henry $1 for each head that appears on two tosses of the coin. What is the opportunity worth to him, in other words what must he pay me to make this two-toss game fair? Let us compute his expectation according to the above principle, which states that his total expectation is the sum of his expectations on each of the tosses. His probability of throwing heads on each toss is ½, and his gain is $1; his expectation on each toss is therefore $.50, and his total expectation is $1. This is the amount that Henry must pay me if the game is to be fair.

We can verify this result, and the principle on which it is based along with it, by obtaining it in another manner. Henry is certain to toss two heads or one heads or no heads, and these are three mutually exclusive events; their probabilities therefore add up to 1. His probability of tossing two heads is ¼, and if he does so he wins $2; his expectation on this throw is thus $.50. His probability of tossing no heads, which means that he tosses two tails, is ¼, but he wins nothing. The remaining probability, that of tossing exactly one head, is found by adding together the other two and subtracting the sum from 1; this gives ½. The corresponding expectation is $.50, so that Henry's total expectation is $.50 plus $.50, or $1, as before. Note how much more easily this result is obtained by the previous method.

This idea of expectation, which has just been introduced, is of great value as a measure of monetary risks of all sorts, whether these risks have to do with betting or not. Although every bet is a risk, obviously not every risk is a bet. Every risk, bet or no bet, is best handled by this same machinery of computing the expectation of profit and the expectation of loss. If an enterprise involves, on the one hand, a risk of loss, on the other a possibility of profit, a comparison of the expectation in the two cases shows at once whether in the long run a profit should be expected. In the absence of other important factors, impairment of capital, legal or ethical considerations, and so forth, the answer will determine the procedure in the individual case. In fact, the introduction of the expectation reduces the question to one of simple accounting.

Let us now leave betting and consider one or two simple illustrations of the application of the idea of expectation to other types of risks:

A shipper of fruit is warned by the weather bureau to protect his shipment for a temperature of 20° F. He has found out from long and careful observation that under these conditions the chance of encountering a temperature of 15° is 1 in 5. He also knows from experience that a temperature of 15° will result in a total loss of 15 per cent of the fruit. To protect for the lower temperature involves an *increased* protection cost of 4 cents on every $1 of profit. For which temperature should he protect?

On an amount of fruit that represents a profit of $1, his expectation of loss, if he elects the 20° protection, is $\frac{1}{5}$ times .15 times $1, which is equal to 3 cents. If he chooses the 15° protection, the loss is exactly the added cost of protection, which is 4 cents. Therefore in this example the extra protection results in a 1 per cent decrease in net profit and, other things being equal, the shipper would protect for 20°.

In practice, under these conditions, the shipper might well decide that it was worth his while to use the extra protection. This would be the case, for example, if he felt that the effect on the consignee of receiving one shipment in five containing 15 per cent damaged fruit, although the latter does not pay for it, was more injurious to his business than the loss of 1 per cent of his profits.

Like ordinary accounting, these probability considerations cannot do more than make a clear quantitative statement to the businessman; and the scope of the conclusions is always strictly limited by the scope of the elements that went into the formulation of the problem. You cannot set out with facts, no matter how correct, concerning automobile fatalities, and end up with conclusions about the death rate from cancer. Probability and statistics may serve business with the same success that they have already served several branches of science, but they cannot replace judgment in the former, any more than they have in the latter, and any attempt to use them in such a way is certain to be disastrous.

Another illustration of the principles under discussion comes from the field of insurance. As the solution to the problem involved is of direct practical interest to almost every individual and business concern, we shall discuss it in some detail.

The theory underlying all forms of insurance can be stated simply as follows: Each of a large number of individuals contributes a relatively small amount to a fund which is paid out in relatively large amounts to those contributors who are victims of the contingency against which they are insured. In the case of life insurance it is of course not the contributor, or insured, to whom the payment is made, but his beneficiaries. We have referred only to individuals; let us agree that in this discussion the word will include groups of individuals or organizations.

The key words in the above statement are, in order, "large," "relatively small," and "relatively large." A little thought will convince one that each of these words is essential to a sound scheme of insurance. It is essential that there be a large number of contributors so that the insurance company can meet all claims and remain solvent. If each claimant did not receive a relatively large payment, as compared to his annual contribution, or premium, there would be no point whatever in insurance, and no sane person would buy it.

Let us see how the insurance company determines the amount of your annual premium that will insure you against the risk of some specific event. The first thing the insurance actuaries must do is determine the chance that this event will take place during the next year. This is done by a study of statistics bearing on similar risks over as long a period as possible, and may be a complicated business. We shall be interested in this aspect of the problem when we discuss the science of statistics; at present our interest is in the application of the idea of expectation of profit or loss. In any event, the net result is a definite numerical estimate of the probability in question; let us say that it is 1 chance in 10, or 10 per cent. To be explicit we shall assume that the amount of your insurance policy is $1,000. Then your expectation of loss is 10 per cent of $1,000, or $100. This is the monetary measure of your annual risk. If the insurance company had no selling or other operating expenses of any sort, and did not wish to operate at a profit, this is what your premium would be. In practice, of course, the premium is "loaded" to cover operating expenses, reserves, and profits, if you are dealing with a commercial company. This loading will usually amount to $30 or more; to be explicit let us say that it is $30, so that your total premium is $130.

You are paying $130 for $100 worth of risk. Does that mean that you are making a bad buy? It does not follow at all. In fact, the interesting question is this: Under what circumstances is it advantageous to pay $(R + x)$ dollars for R dollars' worth of risk? The answer is that it is highly advantageous, even when the extra cost, which we have called x, is considerable, if the contingency against which you are insuring yourself represents a serious financial disaster to you or your dependents. In the case of life insurance you are protecting your dependents against the losses of various kinds consequent on your death, losses that might otherwise prove seriously embarrassing. Thus life insurance has done an immense amount of good, and the fact that you are necessarily paying more for the protection against risk than it is technically worth is of no importance. The same is true when you insure yourself against any sort of financial loss that would be either crippling or embarrassing. This is even more evident when we consider the entire group of insured individuals. Many of the risks insured against represent very rare events. Only a tiny minority of the population, for example, are permanently disabled, in the sense of the insurance companies. But to those who took insurance against the risk and are struck down by this disaster, the relatively negligible amounts they have paid for this protection are the best investments of their lives.

When the contingency against which we insure ourselves represents not a major but merely a minor financial reverse, the case may be otherwise. If an individual can well afford to finance his own small losses, he is very foolish indeed to take out insurance to cover them. He is paying for the protection at least 30 per cent more than the risk is worth, and the insurance company cannot be criticized for collecting the larger amount, since no company can operate without incurring expense. Furthermore, it is not in the sphere of the insurance company to know whether a given person can or cannot afford to take certain losses. That is a decision that he must make for himself.

There are, however, important exceptions to the conclusions of the last paragraph. In the case of certain contingencies the insurance company does a great deal more than merely reimburse the insured. In liability insurance, for example, the amount of the loss is rarely determinable at the time of the accident; it is established later either by agreement or by court decision. It is a part of the insurance agreement that the company take over immediately after the accident has occurred, an arrangement of great value to both parties. For the insurance company is far better equipped from every point of view to handle such matters efficiently. So much so that even if the individual can afford to finance his own liability insurance, he would be very foolish to do so. And it would probably cost him considerably more in the end, as he would have to fight the case as an individual.

In these applications to practical affairs we must not forget that neither the theory of probability nor any other scientific theory works automatically. If we are not skillful enough to use them soundly, taking account of every pertinent circumstance, we can expect nothing from them.

In what we have said of insurance we have of course assumed that if the insured suffers a loss the insurance company will conduct itself in a thoroughly ethical manner. Unfortunately, there are cases where this is not so, even among companies of long-standing national repute. Such cases are of course rare—otherwise there would be no insurance business—but they are nevertheless inexcusable. To protect himself the prospective policyholder should read and understand the entire policy, including the "fine print," before accepting it. If some of the conditions are unclear or ambiguous, he would do well to consult an attorney. Many states have conferred on their Commissioner of Insurance the necessary authority to intervene in such cases. If so, this constitutes an important source of protection.

In our discussion of parts of the theory of insurance we have taken the point of view of the individual. Our conclusions apply even more cogently to many businesses, especially those in a strong financial position. Many such businesses, faced with rising operating costs and declining profit margins, have turned their eyes inward in an effort to find compensating economies; yet they are spending relatively large sums to insure themselves against a very large list of minor losses. Other companies, realizing the waste involved, have started bravely on a program of self-insurance, only to gag at the first uninsured losses. In this matter one must have a long-term point of view and have faith in the law of averages.

CHAPTER VIII

Who Is Going to Win?

THE population of the gambling casinos of the world is largely composed of two classes: those who risk their money for the pleasure and excitement of the play and those whose purpose and hope is to end the play with a profit. As for the true professional gambler, he often prefers to play from the bank's side of the table, where the chances of the game are in his favor, so that he may look to the law of averages to yield him a profit in the long run.

Concerning the members of the first of these classes we have nothing to say. Their winnings and losings are subject to the same laws which apply to the second group, but when the losses outbalance the winnings, they are willing to consider the difference as the price of an evening's entertainment (along with the admission fee, donations to the croupiers, and the cloak-room charge). As for the second class, where no extraneous considerations enter in, that is a different matter altogether. They have set themselves a very difficult problem, namely to win money consistently in a game in which the chances are slightly against them. In case the game is equal for both parties or such that the players play against each other, the house gets its profit in the form of a percentage of the total stakes, or in some similar manner. In the last analysis all these methods are equivalent.

If it is possible for one person to win consistently from the casino, why is it not possible for another to do so, and in the end why not everyone? The answer given by those who believe in the existence of a group able to accomplish this result is that if everyone were to adopt the same "serious" style of game that they play, the losses of the casino would mount so rapidly that it would soon be forced to discontinue. By "serious" play they mean systematic play of the sort discussed in a previous chapter. This theory leads to the startling conclusion that, looking at the matter from the point of view of the casino, the sure way to lose money is to play at games with the odds in your favor. It would follow that the way to win is to give the advantage to your opponents.

In reality there are at least two apparent disadvantages that the bank carries as an offset to the very obvious advantage of the favorable chances of the game. First, the player has the choice of plays and, within limits, of amounts. This point has already been touched on. The choice of plays has no effect in the long run, but the choice of the size of the bet would have an effect on the solvency of the bank, if it were not strictly limited. In practice, the limit in question is so small compared to the resources of the bank that, as we shall see, the effect is negligible. Second, the resources of the bank, though large compared to the effective capital of any one player, are very small compared to those of the public as a whole. The bank is in effect playing against a player who is "infinitely" rich. This brings us to one of the most important problems concerning all forms of risk involving future uncertainties, and the consequences of its solution are illuminating in many directions.

The problem loses nothing by being reduced to the simplest possible form. A and B are playing at heads or tails, or at any game in which their chances are even. The stakes are $1; at each toss the winner receives $1 from the loser. A has a certain amount of money; B has a certain amount. If either player wins all his opponent's money, we shall say that the latter is "ruined." Call the amount of A's capital a dollars, B's capital b dollars. We wish to know the probability that A will succeed in ruining B (or that B will ruin A). When B has lost all of his *b* dollars it means that A has won *b* tosses more than B has won, while at no previous point in the game was B as many as *a* tosses ahead of A.

In our previous problems centering in games of chance we were able to make a list, or imagine that a list was made, of all the equally likely possibilities favorable to an event, and another containing those unfavorable to it. The probability of the event was then found at once by counting the number of entries in the first list and dividing by the total number in both lists. In the problem that occupies us now, no such simple procedure is possible, for the reason that we have no idea in advance as to the length of the game. In fact, there is no assurance that it will ever be finished. Suppose, for example, that A's capital is $200,000, B's $100,000. For A to win he must at some stage of the game be one hundred thousand tosses ahead of B; for B to win he must at some time be two hundred thousand tosses ahead of A.

It is certainly not evident that either one of these two things will happen, no matter how long play is continued. I imagine that the majority of those encountering this problem for the first time are inclined to assign a rather large probability to the third alternative, namely, that the game is never finished. The solution of the problem shows, however, that *if the game is continued long enough,* one or other of the first two alternatives is *certain* to happen. The probability that the game will never be finished is 0, regardless of the amounts of capital of the players. It is to be noted particularly that this astonishing result is a consequence of the solution, and that it was in no way assumed in the statement of the problem.

Since we are unable to enumerate the possible cases, while our definition of the idea of probability apparently requires such an enumeration, how are we to proceed to find a solution? Up to this point we have come across no such intractable problem, for the reason that we have carefully selected problems that can be subdued with a little ordinary arithmetic plus elementary logic. In the present case more of the resources of mathematics must be drawn upon. There is no help for it. We shall have to plunge through it, for to omit taking account of the conclusions involved in this problem would be to impair our understanding of the subject as a whole.

Although it is impossible, except for readers with some knowledge of mathematics, to follow the solution in detail, it may be of interest to see what form the attack on such a problem takes. In any event, the reader who so wishes can conveniently skip the following three paragraphs.

We want to find the probability that A, with a capital of a dollars, will win B's entire capital of b dollars. Imagine that A has succeeded in winning all of B's capital except $1. Then he has an immediate probability of ½ of winning that last dollar on the next toss. But that is not all. Suppose he loses the next toss, so that B has $2. A can still win, and in many different ways. It is only necessary that A win the next two tosses, or three out of the next four, or four out of the next six, and so on. Therefore the probability that A will win when B has a capital of $1 is ½ plus one half times the probability that A wins when B has a capital of $2. The latter probability remains undetermined.

Suppose now that B has exactly $2. As before, there are two possibilities on the next toss, each with probability ½; A either wins or loses. If he wins, B has left exactly $1, which case was just considered. If he loses, B's capital becomes $3. It follows that the probability that A will win when B's capital is $2 is one half times the probability when B has $1 plus one half times the probability when B has $3. Making use of our previous result this last becomes: The probability of A's winning, when B has $3, is equal to three times the probability when B has $1, minus 2. Exactly similarly, the probability for A to win when B has any larger amount of capital can be expressed in terms of the probability when B has $1. To determine the latter there is one more relation furnished by the problem itself. If B has in his hands all of the capital, both his own and A's, the probability for A to win is evidently 0; A is already ruined. To use this condition it is necessary to construct from the above partial steps a general expression for the probability for A to win when B has, say, n dollars, in terms of this amount n and of the probability for A to win when B has exactly $1.

This may be done by either of two methods, one general, applying to all sets of equations of the type we have found, the other special to the problem in hand. The former is very similar to that familiar to all engineers in solving the type of equations known as linear differential equations with constant coefficients. When this expression is found we make n equal a plus b (total capital of A and B), so that the right-hand side of the equation must be equal to 0. This determines the probability that A will win when B has $1, and therefore all of the others. The solution is complete.

The actual result is extremely simple. If A has a dollars and B has b dollars, the probability of B's ruin is a divided by a plus b. The probability of A's ruin is b divided by a plus b. The probability that either A or B will be ruined is found by adding the two probabilities together, since one event precludes the other. The sum in question is 1, verifying the statement made above that the ruin of one or the other is certain if the game is continued indefinitely. Suppose now that one of the players, say A, has enormously more capital than B. Then the probability of B's ruin, which is a divided by a plus b, is very nearly equal to 1, and the larger A's capital compared to B's, the nearer it is to 1, and the more certain is B's ultimate ruin.

In terms of betting odds these conclusions become as follows: If two players sit down to an equitable game of chance, the stakes being the same on each round, and if the first has ten times the available capital of the second, then the odds are 10 to 1 that the second player will be ruined before the first. This illustrates the overwhelming effect of the amount of capital at the player's disposal when the play is prolonged. Suppose that A, with his capital of a dollars, plays fifty rounds each day against a different opponent, each of whom has an amount of capital comparable to his own. The play is assumed to be fair. A is in effect playing a single game against an adversary who is immensely rich, and if he continues, his ruin becomes certain.

A word may be necessary here to avoid a possible misunderstanding. We have shown that if two players, each with a stated amount of capital, play a game of pure chance in which their chances are even, and if the game is carried on for a sufficiently long time, one or the other will lose all his available capital. This conclusion is theoretical, as distinguished from practical, by which we mean simply this: It is certain that if the game is sufficiently prolonged, one or other of the players will be ruined, but in practice this might be a case similar to that of the court sentence imposed on the pig in *The Hunting of the Snark*. This sentence, one recalls, had not the slightest effect, as the pig had been dead many years. Similarly here, if the ratio of the players' capital to the amount risked per game is large, one of the players might well be dead long before his ruin, as predicted in the mathematical theorem, could overtake him. This would surely be the case, for example, if Messrs. Rockefeller and Carnegie had spent their mature lives in matching pennies.

If, on the other hand, the amount risked per game is not too small a fraction of the player's total capital, and if the period of play is moderately long, the predictions of the theorem become an intensely practical matter. This is likewise true in the case of the player, just referred to, who plays against an ever-changing group of opponents.

Imagine ten players in a room playing under the following conditions: Two of them, selected by chance, play against each other until one or the other is ruined. The winner takes on a third party, also selected by chance, and so on. The result is certain to be that after a sufficient period one of the ten has in his possession the entire capital of the other nine, as well as his own. For each individual game must end in the ruin of one or other of the participants, as we saw above.

Suppose that each of these ten players has the same amount of capital. Has the player who enters the game after several rounds have been played greater or less chances of being the final survivor than the two who commence? Consider one of the latter. His probability of winning the first round is $\frac{1}{2}$. If he wins, his capital becomes twice what it was, say $2c$ dollars, and his opponent has only half of this capital or c dollars. According to our general result, the probability of the latter's ruin is c divided by $3c$, which is $\frac{1}{3}$. The first player's chance of winning the second round is therefore 1 minus $\frac{1}{3}$, or $\frac{2}{3}$, assuming that he has won the first round. If he wins both times, his chance of success on the third round is $\frac{3}{4}$, for he now has a capital of $3c$ dollars, his opponent always c dollars. Similarly for later rounds. If he wins the first n rounds, his probability of success on the next is $(n + 1)/(n + 2)$. The probability that he will win all nine rounds is therefore $\frac{1}{2}$ times $\frac{2}{3}$ times $\frac{3}{4}$ times . . . times $\frac{9}{10}$. The product of these nine fractions is equal to $\frac{1}{10}$, for each denominator, except the last, cancels the next following numerator.

Now consider the player who enters the game at the second round. He must play against an opponent who has already won once and who therefore has twice the capital that he has. His probability of winning the round is $\frac{1}{3}$. If he does win, his capital becomes $3c$ dollars, exactly as though he had entered and won the first round, and his probabilities of winning the successive rounds are identical with those computed above. His probability of winning all eight rounds is therefore $\frac{1}{3}$ times $\frac{3}{4}$ times $\frac{4}{5}$ times . . . times $\frac{9}{10}$, and this product is also equal to $\frac{1}{10}$. The same is true of all the players, no matter at what round they enter. A player has the same chance to win, curiously enough, no matter how he comes out in the initial draw for places.

In obtaining the fundamental result that the probability of B's ruin is the amount of A's capital divided by the total capital of the two, it was assumed that $1 changed hands at each round. If this amount is changed to some other, say $5, it is necessary only to divide the capital of each player by 5, and to consider the stakes to be one unit (of $5). Then every step in the reasoning remains as before. Also the probabilities remain the same, as the value of a fraction does not change when numerator and denominator are divided by the same number.

Up to this point we have assumed that A and B are playing an even game, each having 1 chance in 2 of winning each round. When the chances are in favor of one of the players, even by a small margin, the conclusions are considerably changed. In the problem as outlined above, it is almost as easy to consider the chances uneven as even, and it is in this more general form that the problem was solved in the first place. We shall content ourselves with stating the result, which is not quite so simple as in the other case. Let p be the probability that A will win a round; then B's probability is $1-p$. The odds in favor of A on each round are p divided by $1-p$, which we shall call d. Then the probability that B will be ruined, if d is greater than 1, which means that the game is favorable to A, comes out to be

$$\frac{d^a d^b - d^b}{d^a d^b - 1}$$

In this expression a and b have the same meanings as previously, namely, the respective amounts of capital of A and B.

To illustrate the use of this formula we can take a very simple case. Suppose that the odds are 2 to 1 in favor of A, so that d is equal to 2, and that A's capital is $2, B's $3, the stakes being $1. Then the above expression becomes

$$\frac{(4 \times 8) - 8}{(4 \times 8) - 1} = \frac{24}{31}$$

This is the probability that B will be ruined sooner or later.

To find the probability that A will be ruined it is necessary only to exchange the places of a and b, also of p and $1-p$, in the above expression. (d becomes $1/d$.) It comes out that also in this case the probability that either A or B will be ruined is equal to 1. The game always ends. Knowing this we can say that the chance of A's ruin, in the above numerical illustration, is $\frac{7}{31}$. If B's capital is taken to be equal to A's at $2, the corresponding results are $\frac{4}{5}$ and $\frac{1}{5}$. We see that when the odds are against B on each play, an increase in his capital does not help him much. In this example a 50 per cent increase in B's capital has decreased the chance of his ruin exactly 4/155. This is the case of a player (B) against a bank (A), the chances of the game being in favor of the bank.

If the bank's capital is very large compared to that of the player, a is very large compared to b, and since the chances are in favor of the bank, so that d is greater than 1, the term $d^a d^b$ in the numerator of the expression for the probability of B's ruin is enormously greater than the term d^b. The probability of B's ruin is for all practical purposes equal to 1. It is certain. This result applies as strictly to the case of the player who plays only the slowest and most conservative games, such as consistently small bets on the "even chances" at roulette or *trente et quarante*, as it does to his more reckless cousin. Its consummation is only a matter of time.

What about the bank? The bank is playing against the public, and no matter how adequate its working capital may seem to be, it is very small compared to the combined resources of the players. The probability that the bank (A) will be ruined is obtained by subtracting the probability of B's ruin from 1. The result is

$$\frac{d^b - 1}{d^a d^b - 1}$$

This time b is very large compared to a. The expression can equally well be written

$$\frac{1 - 1/d^b}{d^a - 1/d^b}$$

The odds of the game are in favor of the bank, so that d is greater than one. The second term in both numerator and denominator is so small that it can be omitted without appreciable error, and the larger B's capital is, the smaller this term becomes. The probability of the bank's ruin is therefore $1/d^a$. In practice this amount is inconceivably small. On the even chances of roulette, where the bank's advantage is slight, the odds in favor of the bank are approximately 1.03 to 1. Suppose that all the players placed bets on these chances, each bet being 10,000 francs, and that the cash resources of the casino are assumed to be only two thousand times this amount, or 20,000,000 francs. Then the probability that the bank will eventually be ruined is $1/(1.03)^{2000}$, or 1 chance in something like 48,000,000,000,000,000,000,000,000,000.

Small as this chance is, it overstates by far the actual risk taken by the bank, for the following reason: When several players place bets on red and black in roulette, the total amount that the bank can lose on the play is the *difference* between the amounts staked on the two colors. If these amounts were exactly equal on each spin of the wheel, the bank is clearly taking no risk whatever; its losses on one side of the table are paid from its winnings on the other side, while the effect of the 0 is a win from both sides. To make our previous calculation correct, then, we must assume that the *net* amount staked on the opposite chances on each play is 10,000 francs.

The reader who has followed the reasoning of this chapter may have a question to ask at this point. We have seen above that the ruin of either the bank or its opponent is *certain* in the end. If the chance of the bank's ruin is so fantastically small, then the ruin of the bank's opponent, which is the public, is correspondingly certain, and this result is absurd. The answer is this: The bank is playing against an opponent, the public, whose resources are so enormously greater than those of the bank, that we could take the former to be indefinitely great, in other words infinite, without committing an error large enough to be visible. And it does not matter at what fixed rate you win from an opponent who is "infinitely rich," you cannot, in a finite time, exhaust his resources. The bank is in reality playing against a very small and ever-changing group of individuals. While their resources at any given time are strictly limited, their capital is being constantly replenished by the addition of new members; the effect in the long run is the same as though this tiny group had unlimited resources. But in our statement of the problem we assumed that each player has a definite, fixed amount of available capital. These amounts were denoted by a and b. It is entirely correct that if a group of individuals, with definitely limited capital, played *long enough* against the bank, the latter would win *all* of their capital. For another thing, the money won by the bank is not hoarded, but is put back into circulation through the ordinary economic channels, and this fact also was not taken account of in our original statement of the problem. Gambling neither creates nor destroys wealth; it affects only its circulation.

If we leave gambling and turn to businesses which involve future risk, the situation is somewhat similar, but very much more complex. The business house corresponds to the "bank," its clientele and those with whom it does business to the players. Here, too, the latter form a small group whose resources at any time are limited, but which has in effect unlimited resources in the long run, always provided that this group does not melt away. Whenever a future risk is in question there are two fundamental matters to be considered, the expectation of profit (or loss), and the amount of capital. We have been able to obtain some idea of their relative importance under different conditions. In business the future risks are not, of course, determined by a set of rules, as in a game of chance, but are in almost constant fluctuation. These risks include such things as losses from theft, losses from disasters like floods and earthquakes, losses in inventories due to unexpected collapse of prices, failure of sources of supply, labor troubles, and many others. Protection against some of these risks can, of course, be obtained by insurance.

As long as the chances remain favorable to the business house, no matter by how slim a margin, we have seen that the chance of ultimate ruin is small, at least if the fraction of the total available assets risked in any one venture is small. This is an example of the well-known principle of "division of risks," taken account of by the insurance companies. We have also seen, on the contrary, that when the risks become even slightly unfavorable, the amount of available capital, less important in the preceding case, becomes all-important, and that no fixed amount of capital will save the situation in the long run, if there continues to be an expectation of loss. This means that even though a business is conducting its normal, routine selling operations at a profit, and continues to do so, it may be exposed to extraneous risks, of the types enumerated, large enough to change its expectation of profit into an expectation of loss. In this event large amounts of working capital are required, until the situation can be remedied. It is evidently a matter of great difficulty to obtain a reliable numerical estimate of risks of this sort, to which practically all businesses are exposed, and the ability to make such estimates, to within a reasonable margin of error, is an almost indispensable attribute of the successful businessman.

In new businesses the considerations of this chapter are most important. It is difficult enough in an established business to estimate what we have called the *extraneous risks*. In a new business, where experience is entirely lacking, it is doubly so. But that is not all. The lack of experience is even more critical in regard to the ordinary risks of the business. Practically every new business starts out at a loss and remains in that unpleasant position until it can build sufficient sales volume to carry its overhead, in other words until it reaches the "break-even point." The critical question is whether it has enough working capital to survive this period. Thousands of worthy young businesses have fallen by the wayside because their estimates of the required working capital were inaccurate. Frequently this inaccuracy is due to a failure to allow for chance factors, and a very large allowance may be required. A friend, head of a highly successful company which he founded some twenty years ago, once gave me his formula for new businesses. Compute the required working capital based on the most conservative estimates possible, he said, and then multiply by three. This formula is rarely followed. If it were, the bankruptcy rate would drop sharply, but there would be almost no new businesses.

CHAPTER IX

Chance and Speculation

THE conclusions that we reached in the preceding chapter have a certain application to speculation on the exchange. The causes that underlie the movement of security and commodity prices are exceedingly complex, so much so that it is sometimes impossible to distinguish cause from effect. To the extent that we fail to untangle these causes, and to assign to each price movement one or more definite causes, we are forced to think in terms of chance. It is the same in tossing a penny. We may feel confident that there exist perfectly definite causes that explain why the coin came heads on the last toss, but as we are unable to disentangle them we are forced to explain the toss in terms of chance.

We shall compare the speculator, in several respects, to the participant in a game of chance, in which skill, or judgment, also enters. In making this comparison we are not in search of rules for speculation, for we do not believe in the possibility of exact rules of the sort and are sure that if they existed they would defeat their own purpose by putting an end to all speculation. We are in search, rather, of a more exact formulation of the chances of speculators, in the long run, for profits or losses. We shall discuss here only speculation on the stock market; the same considerations apply to the other exchanges.

Before proceeding, however, we must review rapidly the elementary facts about the stock market. In doing so, we shall confine our discussion as closely as possible to the buying and selling of stocks for speculative, as distinguished from investment, purposes.

This distinction cannot be made sharply in all cases; the purchase of stocks frequently partakes of the characteristics of both. Without quibbling over definitions we shall say that when the long-period yield of the stock in any form (cash or stock dividends or rights), or when the anticipated yield, is a major consideration in the purchase, the stock is bought for investment purposes. When the yield of the stock is at most a minor consideration, we shall say that the stock is purchased for speculative purposes. In the transaction of "short selling," since the sale precedes the corresponding purchase, there can be no question of yield; all short selling is speculative in character.

When you purchase or sell a stock, you place the order with your broker, who charges you, or debits to your account, a certain amount representing his commission. This commission depends on the total value of the shares involved in the transaction. Expressed as a percentage of this total value it may seem small, but it represents, together with interest on borrowed money, the broker's profit on the deal. In speculative accounts, where large blocks of stock are bought and sold within short intervals of time, the monthly commissions may reach a considerable total. The amount of the commissions evidently depends less on the speculator's effective capital than on the rapidity with which he buys and sells. A speculator with a substantial capital on deposit at his broker's, who trades rapidly, may pay many times the commissions paid by another speculator with an effective capital several times as great, who conducts his trading at a more leisurely tempo. A certain speculator, for example, with a capital of several hundred thousand dollars, paid commissions averaging, over a period of years, $20,000 to $30,000 per month.

The usual practice among speculators is to deposit with the broker only a part of the capital necessary to conduct their transactions; in other words, to operate on margin. The amount of this deposit, which may be either cash or acceptable securities, must exceed the prescribed margin requirements at the time in question. This requirement represents a certain fraction of the total value of the stocks purchased. The balance of the required capital is provided by the broker, and the amount of this interest, together with the commissions, must be taken into account by the speculator in estimating his prospects for profits. In practice, speculative accounts may become very involved, and margin requirements are computed in terms of the total debit or credit of the account, and the current value of the shares that it is "long," and of those that it is "short."

The effect of operating on margin is to increase the volume of the speculator's transactions, and therefore to increase his expectation of profits or losses. By the same token, his chance of being wiped out is very much increased and, furthermore, the interest that he must pay makes the odds of the "game" less favorable.

As an offset to commissions and interest, the speculator receives any dividends that may be paid on stocks that he is holding. How important an item this is depends of course on the particular stocks traded in, and should be taken into account by the individual speculator. Its importance is diminished by the fact that dividend prospects cause the price of a stock to rise, other things being equal, before the "of record" dividend date, and to fall immediately after. To the investment buyer, on the other hand, these minor fluctuations in price have little significance. As he has bought the stock for yield, or for long-period enhancement of value, perhaps as a hedge against inflation, a temporary decline in price represents merely a paper loss.

The situation of the speculator is in certain important ways analogous to that of the participant in a game of chance in which the players play against each other, and in which the gambling house, or the banker, in return for supplying the facilities for play, charges an amount proportional to the total stakes placed. The stockbroker corresponds to the "banker" in the game of chance, and has much more right to the title "banker" than does the latter, since a part of his business consists in loaning money to his customers. The other speculators in the stock issues in question correspond to the other players in the game of chance. The commissions and interest that the speculator pays to the broker correspond to the percentage of stakes or bets paid by the player to the "bank."

If the speculator is a professional trader or a floor specialist, a member of the stock exchange who buys and sells from the floor of the exchange on his own account, these remarks have to be revised. Instead of commissions, the amount that he pays per year for the privilege of trading is the interest on his investment in a membership to the exchange, together with his annual dues.

If we compare the speculator in certain respects to the participant in a game of chance of a particular variety, we must at the same time keep in mind the many and important differences between the two situations. In the first place, the buyer of common stocks, regardless of what his individual point of view may be, is dealing in equities in corporations, which presumably have a value related to their probable future earnings. He is, in fact, trading in these values, regardless of the fact that he may buy the stock of the X company, not because he believes that the profit prospects of this company are good, or that the trend of the market should be upward, but because he believes that others believe it. Similarly, he may sell the shares of a certain company short because he believes that they are selling higher than the opinions of other traders warrant, although he himself believes that the prospects of the company justify, eventually, a much higher value.

This constitutes what might be called a "principle of momentum." It means that a market on the upswing tends to go higher until a new set of forces comes into play, while a market on the downswing tends to go still lower. There is no parallel to this aspect of the situation in the games of chance which we have been discussing.

In the second place, in a game of chance like poker the total winnings of those players who finish with a profit is strictly equal to the total loss of the remaining players, leaving out of account the percentage paid to the bank. In the case of stock dealings, however, this is not generally true. In a rising market the group of investors and speculators, as a whole, shows a profit; in a falling market this group, as a whole, shows a loss. These profits and losses correspond to the change in appraisal value of all stocks. If one hundred men own equal shares in a building appraised at $10,000,000, and the appraisal value drops to $8,000,000, each shows, in his list of holdings, a loss of $20,000. There are no corresponding profits on the part of other investors.

So when the price level of stocks is fluctuating, profits and losses of traders cannot be said to balance each other like those of a group playing a game of chance. In a long period, however, the change in the level of all stocks is not large; furthermore, we are discussing here only that part of the total transactions that can be classified as speculative, so that the total profit or loss of the total group of speculators can be assumed to be small. This conclusion is entirely independent of the amount of short selling that has taken place during the interval in question; this amount does not affect the accounting at all.

If we assumed that the price level of all stocks is slowly rising, over long periods, it would be possible for this factor, along with dividends, to offset the effect of brokerage commissions and interest, so that speculators, as a whole, would be playing a game with the odds even, as in a private game of poker, as distinguished from one at a gambling house. On the other hand, the individuals owning the stocks in question at the end of the period are left "holding the bag." In what follows we shall not make this assumption of a rising price level; instead we shall assume that over the long period in question speculators, as a group, show no profit or loss.

In order to apply the conclusions of the preceding chapter to the situation before us, it will be necessary to consider successively several simple situations relative to games of chance, before arriving at the one that interests us. We have seen that in a game of pure chance the rules of the game constitute the only pertinent knowledge, that therefore each player has the same amount of knowledge and hence the same chances as any other player. If a group of players play a game of pure chance against each other, the bank not taking part but merely collecting its percentage of bets placed, the players are on an equal footing with each other, but collectively they are losing to the bank.

We know, too, from the data of the previous chapter that if the game is continued indefinitely, regardless of the amounts of capital at the players' disposal, all but one of them will eventually be "ruined." The survivor will leave the table with the entire capital of the group, less the commissions taken by the bank. If the group of players is continually changing, some players dropping out and others taking their places, this conclusion must be modified. We can then say that if the game is continued long enough, all of the original participants will be "ruined" (or will drop out for some other reason) and their places will be taken by others. And this conclusion applies equally well to the group present at the table at any particular time, no matter how long the game has already been in progress. For the odds of the game are against each of the players, due to the commissions paid to the bank.

If the game is one that involves skill as well as chance, the players will not be on an equal footing with each other. It is important to notice for our purposes that it is the *relative* skill of the players that matters, for they are playing against each other. They may all be masters at the game, or they may be beginners; but if their skill is equal, their chances are equal.

It is to a game of the latter sort, where skill and judgment are contributing elements, that speculating on the exchange most nearly corresponds. In speculating in stocks there is a large element of chance, never to be left out of account. But you are playing a game against other speculators each subject to the caprices of fortune, just as you are, and whether the odds of the game are with you or against you depends most of all on your knowledge, judgment, and skill as compared to the extent to which these qualities are present in the group of speculators against whom you are playing. This is a group of which you can know very little, and one whose membership is changing from day to day.

In addition, there is the matter of available capital. We saw in the preceding chapter that when the odds of the game are in your favor, and when your capital is moderately large compared to the amount risked at any one time, the chance of your ultimate ruin is small. On the other hand, if the odds of the game are even slightly against you, a large amount of capital does not prevent your probability of ruin from being large.

Naturally enough, the larger this capital, the longer its life; if it is very large compared to the amounts risked, it might well require more than a lifetime to exhaust it. But even so, the prospect, when the odds are against you, is not enticing. The question of whether to speculate or not to speculate (looked at solely from the point of view of profit and loss) boils down to your judgment as to whether you possess the requisite qualities in the required degree. If the group against whom you are playing contains many individuals with large capital and "inside" information concerning the stocks in which you are dealing and, on the other hand, you have neither a large capital nor "inside" information, the odds of the game are decidedly against you, and you are certain to be wiped out if you continue to speculate long enough.

Roulette is a game of pure chance. The player is permitted to place his bets as he pleases, provided that they are between the limits set by the bank. He can play a conservative game, or a steep game; he can risk only a small part of his capital at any time, or he can bet up to the limit each time. But his manner of placing bets does not change the odds of the game in the smallest degree, and these odds are always slightly against him because of the advantage taken by the bank in fixing the rules of play.

Poker, on the other hand, is a game of chance and skill. Other things being equal, the more skillful player is certain to win in the long run. When poker is played at a gambling house, the latter takes its commission from each pot that is won; this is an illustration of the sort of game to which we have compared speculating in stocks. In the poker game you are pitting your skill against that of the other players, who correspond to the group of stock speculators. There is this important difference, however; in the game of poker you sit face to face with your antagonists, whose habits and expressions you are free to study. If you have followed the excellent advice of Hoyle, never to play with strangers, you already possess some information concerning your opponents. In speculating, on the other hand, you always play with strangers. You do not meet your antagonists; indeed, you rarely know the identities of any of them.

Although the speculator is thus very much in the dark when it comes to the crucial matter of comparing his own astuteness to that of the group interested in the same stock, there are a few things that he does know. He knows, for example, that in boom periods, such as the few years preceding the autumn of 1929, the "public" is in the market. Amateur speculators very much outnumber the professionals; many are "shoestring" operators, whose limited capital makes them particularly vulnerable. The net result is that at such a time the speculator can count on a low order of competition, as compared, for instance, to that met with in markets such as those of 1932, when the law of survival of the fittest operated in its most ruthless manner and few replacements appeared for those who fell by the way-side.

To enumerate the elements of competition that determine the speculator's relative merit, and hence the odds of the "game," would be to list a large number of the factors that affect the market in any way. Those that are most important in determining the odds for or against the speculator are of two kinds: factors that concern the past, and those that concern the future. The former represent knowledge; the latter, judgment. Knowledge of past facts can in turn be split into three classifications: facts concerning the company whose stock issue is in question, facts concerning the market position of the stock, and facts concerning the market as a whole. The element of judgment can be divided into exactly the same three classifications, this time with reference to future trends instead of to past facts.

Let us pass these three classifications in rapid review. The first contains such things as knowledge of the company's product, its sales prospects, its management, its stock issues, its debt structure, its current financial condition, and so on. The second has to do with the market situation of the stock, whether any groups or "pools" are operating in it, the extent of the short interest, the floating supply of the stock, orders above and below the market, and other things. The third concerns the market situation as a whole, past and future, what forces are agitating it or are likely to agitate it, business prospects, economic trends, the likelihood of significant legislative or international developments, and many other factors.

The three classifications in this analysis are by no means of equal importance. In general it is safe to say that sound common stocks move up and down together, constituting what is known as the major trends of the market. It is these swings that are measured by the Dow-Jones averages and that are considered in that widely known empirical theory called the Dow Theory. It is possible for the speculator, by diversifying his interests, to eliminate, to a considerable degree, all considerations except those affecting the market as a whole. This would be largely true as applied even to one or a few sound stocks, were it not for the fact that individual stocks can be manipulated, to the advantage of a small number of operators at the expense of the public, and that professional traders and floor specialists have access to facts that give them a pronounced advantage over other speculators. In addition, when an individual stock does move against the general market trend, there is always a good reason for it and this reason is known in certain cases, to those close to the company long before it is known to the public. In other words, there are certain privileged classes of speculators, for whom the odds of the "game" are more favorable than they are for others.

We are not concerned here with the ethics of speculation. It is certainly true that all methods of manipulation that depend on circulating false reports or rumors of any nature are thoroughly dishonest. In our analogy with games of chance they are to be compared to cheating at cards or using loaded dice. If you are speculating in a stock that is being manipulated in this manner, you are in the position of an honest player in a dishonest game. Until the time comes when such practices are effectively banned, speculating in stocks for the individual, no matter how competent he may be in the legitimate phases of trading, will remain subject to very unfair risks.

The conclusions that we reached in the previous chapter, which we are making use of here, apply to the probabilities in the long run. In any activity in which chance plays a prominent part, whether a game, a business, or stock market speculation, the unusual is continually taking place. Because the odds are against you in roulette, it does not follow that you will lose in a limited period of play. What is certain is that if you continue to play long enough, with the odds even slightly against you, you will lose your capital. Similarly, the ill-equipped speculator, with the odds against him, may win over a limited period and even heavily, for the element of chance in the market is very large. On the other hand, the well-equipped speculator, with the odds in his favor, may lose heavily over a limited period. It follows that results of speculation over a short period are not a reliable index of the odds for or against the individual speculator. A series of losses does not necessarily mean that the odds are against him, nor a series of successes that the odds are with him. Short-period results merely create an implication, which the speculator should take account of, along with other things, in deciding whether the odds are with him or against him.

This large element of chance that is always present in stock market trading represents the unpredictable and the unknowable. The full causes that underlie changes in price are complicated beyond the power of man to unravel. Furthermore, as in many games of chance, an insignificant cause may produce effects of major importance. A murder at Sarajevo in the year 1914 led to events that affected the market for decades. Since this is so, it is natural that false and superstitious ideas concerning luck and its relation to the individual are found in the minds of some speculators.

In order to illustrate the conclusions toward which we are progressing, imagine a group of speculators operating in a certain common stock. The membership of this group is continually changing, each member has a certain amount of available capital, and each is equipped with his individual knowledge and judgment. If the price of the stock, over a long period of time, returns to its initial level, then this group, as a whole, is losing, to the extent of brokerage commissions and interest charges. The odds of the game are against the group, as a whole.

But there may be certain individual members of it whose knowledge and judgment are such that the odds of the game favor them. Any such superior trader is in a position similar to that of the roulette banker; the odds of the game favor him in playing against the public, whose resources are enormously greater than his own. If his available capital is large compared to the amount risked at any time, the chance of his losing, in the long run, is very small indeed, as we saw in the last chapter. The speculator, under these conditions, although his results will vary with the swing of chance forces, is nearly certain to win in the long run. We have confined these remarks to operations in a single stock. They apply equally to the speculator operating in many stocks, and in that case it is more reasonable to assume that the price level returns, after a long period, to its initial value.

Such conditions as those discussed here are frequently realized, especially by professional traders, who pay less than others for the privilege of trading and who have access to valuable inside information concerning the market. There are also traders who, in spite of having the odds in their favor, end in disaster. When the facts in such cases are known, it is usually found that the scale of operations was too large for safety, in view of the amount of available capital. If the principle of division of risks is not respected, even having the odds with you is not a sufficient guarantee of success.

The other members of our hypothetical group of speculators, for whom the odds of the game are unfavorable, are in exactly the opposite position. They correspond, not to the banker, but to the players of roulette, in so far as ultimate results are concerned. In the long run, they are certain to lose.

In this brief discussion of the expectation of the speculator, we have grouped together, under the head of chance, all those unpredictable elements of every sort that play so important a part in the market. We have done so for the reason that, to a large extent, these chance factors affect all speculators alike, while we have been in search of those factors that distinguish one speculator from another. It would be very convenient (especially for the authors of theories of how to speculate) if the role of chance in the market were smaller than it is. Then it would be easier to assign a plausible cause to each price fluctuation. But chance and speculation are good traveling companions, and while there can be chance without speculation, there will never be speculation without chance.

CHAPTER X

Poker Chances

ONE of the finest illustrations of the laws of chance is furnished by the game of poker. It is not a game of pure chance, like dice and roulette, but one involving a large element of skill or judgment. The required judgments, however, are based squarely on probabilities, and some knowledge of the chances of the game, however acquired, is essential. Owing to the presence of the element of skill, the study of the probabilities of poker is by no means so simple as in those games which we have previously considered. On the other hand, the probability problems of poker have a richness and breadth of application that are lacking in the other games.

We shall therefore make a more careful study of the chances in poker than of the chances in any other game. Those readers who are not familiar with the game will wish to omit the next two chapters.

The game of poker is distinctly American in character and associations, in spite of its French parentage and its popularity among many peoples, including the Chinese. The game was introduced in America and remodeled before the end of the frontier period, into which it entered most harmoniously; it was in this period that it accumulated its most picturesque tales and traditions.

The fact that few games better illustrate the theory of chance than poker has not been overlooked by writers on the subject. As the cards are dealt, each player has a definite chance, readily computed, to hold each of the significant hands. A table showing the odds against a player's holding each of the hands "pat" (on the deal) is given in Hoyle, and correctly given. The odds or probabilities on incomplete hands, such as four straights and four flushes, are not included. These will be listed here later. In the chapter on poker in Proctor's book, *Chance and Luck,* the same thing is given in another form: a table showing the number of equally likely ways in which "pat" hands can be formed.

A second point where poker probabilities are encountered is in the matter of improving hands on the draw. A player holds a pair of jacks, to which he draws three cards. What are his chances of improving his hand?—of making two pairs or threes and so on? On these questions the majority of the writers on poker have scarcely done themselves or their readers justice. The probabilities in question, or rather the corresponding betting odds, are given *in toto* in Hoyle, but unfortunately less than half of them are accurate, several of the blunders being such as to affect the conclusions as to correct play. Proctor passes over the subject with a few remarks, mentioning casually the chances in two or three of the simpler cases. He further quotes from *The Complete Poker Player* as follows: " 'Your mathematical expectation of improvement is slight' " (in drawing two cards to triplets) " 'being 1 to 23 of a fourth card of the same denomination, and 2 to 23 of another pair. . . .' " The first probability, that of making fours, is correct as given, but the second should be 2 to 29 (or 1 chance in $14\frac{1}{2}$), an error amounting to 14 per cent which Proctor fails to point out. We may notice also that the author of *The Complete Poker Player* does not appear to distinguish between mathematical expectation and probability.

In the article "Poker" in the eleventh edition of the *Encyclopaedia Britannica,* a table of probabilities in drawing to most of the hands is included and it is an improvement over that of Hoyle. Nevertheless the figures given there for all draws involving a single pair are incorrect. In the latest edition of the *Encyclopaedia Britannica* all of these errors are corrected, with one exception—the probability of improvement on a two-card draw to a pair. On the other hand, draws to straights and flushes are correctly treated in all these tables (with a few exceptions), these calculations being very simple and direct. The cases of pairs or triplets evidently deserve particular attention.

Another problem that inevitably comes up, above all in "jack pots," is the effect of the number of players on the probabilities of the game. Every beginner knows that this effect is very pronounced, that a hand with excellent prospects in a three-handed game may be quite worthless with seven or eight players. The result is that the principles of sound play are widely different in the two cases, and less so in the intermediate ones. Before discussing this question it is necessary to have before us the various chances of the individual player.

Discussing the value to the practical poker player of an intimate knowledge of these probabilities, Hoyle says, "The player should . . . have some idea of the chances for and against better combinations being held by other players, and should also know the odds against improving any given combination by drawing to it." Further on we find, "In estimating the value of his hand . . . before the draw, the theory of probabilities is of little or no use, and the calculations will vary with the number of players engaged. For instance, if five are playing, some one should have two pairs every fourth deal, because in four deals twenty hands will be given out. If seven are playing, it is probable that five of them will hold a pair of some kind before the draw."

These last remarks merely prove that the author had vaguely in mind some ideas that may be more or less correct but has failed to express them with sufficient precision for his readers to be in a position to judge. The phrase "It is probable that" implies only that the chance is better than even. What we must know is precisely *how* probable the event in question (five pairs among seven players) is.

As a matter of fact, the statement seems to involve one of the commonest fallacies connected with probabilities, that of assuming that the most probable single event of a series is more likely to happen than not. Let us illustrate. If I toss ten coins simultaneously the possible results are as follows: five heads and five tails, four heads and six tails, six heads and four tails, and so on to ten heads or ten tails. The *most probable* of all these eleven cases is the first one given, five heads and five tails, but the probability of this event is only about $\frac{1}{4}$, so that the odds against the most probable case are 3 to 1. The probability of each of the next two cases is $\frac{1}{5}$, that of the following two $\frac{1}{9}$, the next two $\frac{1}{23}$, the next two $\frac{1}{105}$, and of the last two 1/1048. It is seen that each of these cases is less probable than the first (five heads and five tails), but their sum is greater.

Similarly in the case quoted from Hoyle; the most probable number of pairs in a deal of seven hands is *not* five, as the author of Hoyle seems to have thought, but three, and the chance that exactly three of the seven hands contain exactly one pair (leaving out of account stronger hands) is not far from 1 in $3\frac{1}{2}$. The chance that exactly five hands contain one pair is roughly 1 in 13. The method of making these calculations will be given in the next chapter.

The quotation from Hoyle continues: "Unfortunately, these calculations are not of the slightest practical use to a poker player, because although three of a kind may not be dealt to a player more than once in forty-five times on the average" (forty-seven is the correct figure) "it is quite a common occurrence for two players to have threes dealt to each of them at the same time." If the calculations referred to are not of the slightest use, the reason is not difficult to find. And I must confess my inability to understand the significance of the remark last quoted. It is true that it is not rare for two players to find threes in their hands as dealt; in fact, in a seven-handed game this will happen about once in 117 deals, on the average. But as this fact is supposedly known to the informed player, as well as the first one, why should it make a difference one way or the other?

There can be no doubt that in poker an exact knowledge of the chances is of considerable advantage to the player, at least if he has sufficient judgment to know how to use it effectively. On the other hand, nothing would be farther from the truth than to pretend that familiarity with the chances, no matter how complete, makes a good poker player. The old hands at the game are sometimes inclined to feel that their long experience at the table exempts them from the necessity of taking account of the tables of chances.

Now it *is* perfectly possible to learn the probabilities of the game by watching the fall of the cards, in spite of the fact that the majority of the hands are not shown. To do so is to achieve something of a tour de force both as to accuracy of observation and patience, not to mention the possibility of becoming poor in the interval, but it can be done. After sufficient experience the player has a somewhat vague idea of a set of figures that he could easily have learned in an hour in the first place. He may have become an expert player, but it is certainly not to be attributed to his knowledge of the chances of the game.

The test of the accuracy of the knowledge that he has acquired with so much patient observation is his ability to detect gross errors in the calculated chances. When a great expert on the game states that the chance of improving on a two-card draw to a pair and an odd card is 1 in 5, the correct figures being 1 in 4, an error of 25 per cent, we cannot place too great confidence in the accuracy of his observations. Every poker player knows the particular importance, especially from the point of view of deception, of the case just cited. As the chance of improving a pair on a three-card draw is 1 in 3½, the difference between the two figures given has a great influence on sound play. While no doubt the expert in question plays these hands with great skill, and suffers little or none from his defective knowledge of the chances, the same cannot be said of the inexperienced players whom he is instructing.

If one watches the play of inexperienced players closely, the conviction is rapidly forced on him that even an elementary knowledge of the chances and of probability principles would much improve their game. One sees them make a practice of entering pots where the expectation is against them from the start, and their occasional inevitable wins under such circumstances only encourage them to further unsound methods. One sees them hanging on in the face of almost certain loss, believing that they are somehow "protecting" what they have previously put into the pot.

It is perhaps in connection with these various unsound practices that the theory of probability can be of most aid to the poker player, for every player develops habits which represent his normal procedure; from these he makes departures as his judgments of players and situations require them. As a result one not infrequently comes across players of unusual judgment (*poker sense,* it is called), who handle the more difficult situations with great skill but lose small amounts regularly in the more commonplace hands by one or two unsound practices that could very easily be remedied.

We have been saying a good deal about the probabilities of the game of poker; let us examine a few of the more interesting ones a little closer, beginning with those that have to do with the draw. We recall once again the elementary definition of probability or chance. To find the probability of an event, we list the total possible cases, all equally likely to occur, and count those in the list that are favorable to the occurrence of the event. The probability of the event is the number of favorable cases divided by the total number of cases.

Suppose I hold a four flush, to which I draw one card. What is the chance of filling it? There are forty-seven cards left in the pack, of which nine are of the same suit as the four in my hand. There are thus forty-seven possible distinct results of my one-card draw, no one more likely to occur than another, of which nine are favorable. The probability of filling is therefore $9/47$; the chance is 1 in $5\frac{2}{9}$.

"But," says Mr. Reader, "you talk as though yours were the only hand dealt. Suppose you are in fact the fourth hand in the deal. A good part of the deck has already been dealt out, and two or three players have made their draws ahead of yours. How do you know but what all or most of your suit is in these other hands or in the discard pile? In fact, how can you say what your chances are unless you know how many cards of your suit are left in the pack?" We shall linger awhile over this objection, as it represents the only real stumbling block to an understanding of the chances of the poker draw.

Let us look at the matter in this way: I have seen exactly five of the fifty-two cards in the pack, and I know that four of these are spades, say, and the fifth a diamond which I discard. If I am the first to draw in a five-handed game I shall receive the twenty-sixth card from the top of the deck. I can, then, state the whole problem in this form: What is the chance that the twenty-sixth card in a well-shuffled deck is a spade, knowing that among the top twenty-five cards there are at least four spades and one other that is not a spade? If I did not have this last information, the chance that the twenty-sixth (or any other) card is a spade would be 1 in 4, thirteen favorable cases out of the fifty-two. But I know that this card cannot be one of the four spades held in my hand, nor the discarded diamond. There remain forty-seven possibilities, nine of them spades. I have taken account of the fact that any or all of these nine spades may be included in the twenty cards of the top twenty-five which I have not seen. If one of these twenty cards were accidentally exposed to me, my chance of filling would be changed, *as it depends on the extent of my information*. If the exposed card were a spade, my chance would become $8/46$, which is less; if it were *not* a spade, my chance would become $\frac{9}{46}$, which is more. Now suppose that I am not the first to draw. Then the card I draw will be the thirtieth, or the thirty-second, or the thirty-sixth, or some other. Nothing is changed. Whether it is the twenty-sixth or the thirty-sixth card from the top of the deck has no effect on the argument.

In case any last doubts on this important point remain, it should be added that it is not at all difficult to prove the result in the following way: Suppose for definiteness that my hand makes the first draw, so that the card drawn is the twenty-sixth from the top. We can compute the chance that any number from nine to zero of the missing nine spades are in the balance of the deck, which contains twenty-seven cards in all, and treat the probability that the twenty-sixth card is a spade separately in each case. Later on we shall carry out something similar to this when discussing the chances on the deal. This method removes the objection based on the possibility that some or all of the spades have already been dealt out, but is quite unnecessary when once the point has been grasped that the possible cases involve all cards which have not been seen, whether in the hands, on the table, or even under it, provided only that the game is honest.

Suppose the player discovers, upon examining his hand with the caution recommended by Hoyle, that it contains a four-straight flush open at both ends. What is his chance of improving on the draw (leaving the formation of a pair out of our calculations)? Of the forty-seven cards remaining in the deck, of which the player has no knowledge, there are two that would complete a straight flush, seven others that would complete an ordinary flush, and six others that would give him an ordinary straight. His chance of improvement is therefore $2 + 7 + 6$ divided by 47, or $\frac{15}{47}$, or roughly 1 in 3. The odds against the player are 2 to 1. They are given in Hoyle as 3 to 1, an error of 50 per cent in the odds, $33\frac{1}{3}$ per cent in the chances. The player's chances for a straight flush, a flush, and a straight are respectively $\frac{2}{47}$, $\frac{9}{47}$, and $\frac{8}{47}$.

These draws of one card are the simplest to handle. Two-card draws are well illustrated by draws to a three flush, although this play is used in sound poker only under the most extraordinary circumstances. Take the case where the player's hand contains three spades and two cards of other suits, which are discarded. Forty-seven cards remain in the deck, unknown to the player, of which ten are spades. To find the probability of getting the flush we go back to the definition and list the possible cases. How many are there?

We wish to know how many different hands of two cards each can be dealt from a pack of forty-seven cards. Now there are forty-seven possibilities for the first of the two cards, and for each of these there are forty-six possibilities for the second card, making in all 47 times 46 cases, and clearly they are equally likely. Now we must pick out from these the favorable cases; this means those where both the cards are spades.

This problem is the same as the following: How many hands of two cards each can be dealt from a pack of ten cards? The pack in this case consists entirely of spades. There are ten possibilities for the first of the two cards, and for each of these there are nine for the second, making a total of 10 times 9 favorable cases. The probability of filling the three flush is therefore

$$\frac{10 \times 9}{47 \times 46}, \text{ or } \frac{45}{1,081}.$$

The chance is 1 in 24, odds 23 to 1 against.

In such an enumeration of the possible and of the favorable two-card hands we have considered as distinct two hands containing the same cards but in a different order. By doing this it was possible to say with absolute assurance that each of the possible cases is equally likely. If two hands containing the same cards are considered identical, as they are, of course, in poker, can we still say that the possible cases are all equally likely, so as to be able to use the definition of probability? The answer is that we can, as a little reflection will show. It is in fact usual, in all these problems, to make the analysis in this way, since the numbers involved are then smaller.

In technical language, when the *order* of each arrangement of cards (or in general, of cases) is taken account of, we use the word *permutations*, otherwise the word *combinations*. In the above calculation of the chance of filling a three flush the permutations two at a time of a deck of forty-seven cards numbered 47×46. How many combinations two at a time are there? Evidently half of the above number; for to each hand such as the 10 of hearts and the 5 of spades there corresponds another containing the same cards in the opposite order, namely the 5 of spades and the 10 of hearts. Similarly for the favorable cases; there are $\frac{1}{2} \times 10 \times 9$ favorable combinations. The value of the probability obtained in this way, $\frac{1}{2} \times 10 \times 9$ divided by $\frac{1}{2} \times 47 \times 46$, is evidently the same, $\frac{1}{24}$.

In probability problems we are constantly required to count the number of favorable cases and the total number of possible cases, and this usually involves enumerating the combinations of a certain number of things taken so many at a time. The rules for calculating the number of these combinations have been worked out once for all, just as in school arithmetic the child is provided with a set of rules covering addition, multiplication, etc., and is not concerned with the origin or correctness of the rules. We shall satisfy ourselves here by merely stating the rules for finding the number of combinations in a given case. Of four distinct letters, A, B, C, D, it is required to find the number of distinct two-combinations. In this simple case we can list them at once; they are, AB, AC, AD, BC, BD,

CD, six in all. It is convenient to abbreviate the expression "the two-combinations of four distinct things" by writing it C_2^4. Then we know that $C_2^4 = 6$. The general rule for calculating these combinations will be made clear by a few numerical illustrations:

$$C_2^4 = \frac{4 \times 3}{2 \times 1} = 6; \quad C_3^6 = \frac{6 \times 5 \times 4}{3 \times 2 \times 1} = 20; \quad C_1^{47} = \frac{47}{1} = 47.$$

$$C_5^{52} = \frac{52 \times 51 \times 50 \times 49 \times 48}{5 \times 4 \times 3 \times 2 \times 1} = 2,598,960.$$

It is seen that there are always the same number of factors in numerator and denominator, and the rule for writing them down is self-evident from these examples. In what follows we shall not need more than this practical rule for computing. The last of these numerical illustrations, C_5^{52}, in other words the five-combinations of fifty-two things, gives the total number of possible, distinct poker hands. Thus there are 2,598,960 poker hands.

There is another method of computing poker probabilities that is convenient in some cases. It consists in considering the chances separately for each card drawn, and in using a rule given in Chapter VI, namely: The probability of an event consisting of two separate events is the probability of the first event multiplied by the probability of the second, the latter being computed on the assumption that the first event has already taken place.

To illustrate this attack let us return to the two-card draw to a three flush. There are ten spades left in a deck of forty-seven cards, so that the probability of drawing a spade on the *first* card drawn is $^{10}\!/_{47}$. Assuming now that this draw has been favorable, there remain nine spades in a deck of forty-six cards, so that, *on this assumption,* the probability of a spade on the second draw is $\frac{9}{46}$, and the probability of filling the three flush is 10×9 divided by 47×46, or $45/1,081$, as before.

We now leave straights and flushes and turn our attention to draws to hands containing a pair, threes, and so on. First of all, let us use once more the latter of the previous methods, applying it this time to computing the chance of making fours when two cards are drawn to threes. Of the forty-seven cards that remain, one only is favorable. Now there are two different ways to make fours: We can draw the winning card on the first draw, or we can draw some other card the first time and win on the second. The probability of drawing the winning card first is $\frac{1}{47}$; when this has taken place, it makes no difference which of the remaining forty-six cards is drawn second. The probability of winning on the first card is therefore $\frac{1}{47}$. There remains the case where we do not draw the favorable card until the second attempt. The probability of *not* drawing it the first time is $\frac{46}{47}$, after which the probability of success on the second draw is $\frac{1}{46}$. The total probability of success either way is therefore $\frac{1}{47} + (\frac{46}{47} \times \frac{1}{46})$, which is equal to $\frac{2}{47}$, or 1 chance in $23\frac{1}{2}$.

The commonest draw in poker is of course to a pair, and for this reason the chances for improvement should be accurately known by every player. Yet it is exactly in this case that the authorities on the game give the most contradictory figures for the chances. If it were not for this situation we should be inclined to pass over the details of the calculations, giving only the results, but under the circumstances it is necessary to demonstrate to those interested in poker chances that the figures presented here are the correct ones.

All good players, when they receive a pair on the deal, take advantage of the opportunity thus afforded to "mask" their hands by varying their draws between three and two cards. The normal draw, which gives the greater chance for improvement, is of course three. The question of how frequently it is advisable to shift to a two-card draw is a very interesting one. Its solution depends on the relative chances of the two draws, the number of players in the game, the position of the player at the table, previous draws by other players, his judgment as to the tactics of others, and various other considerations. We will postpone further discussion until we have before us those factors that are present in every case, and on which the player's decisions in a particular situation must depend.

The first case to be considered is that of three-card draws to the pair. What is the probability of the player's having two pairs after the draw? *When a hand is spoken of as containing two pairs, threes, and so on, it will mean, unless otherwise specified, that it contains no higher count.*

After the deal the player finds in his hand one pair, let us say a pair of aces, and three other cards not containing a pair, let us say a king, a queen, and a jack, so as to make the problem as definite as possible. He discards the last three, and asks for three cards. (It sometimes happens that a player discards the smaller of two pairs; for example if he believes that threes will win but not the two pairs that he holds. In this case the chances on his three-card draw are very slightly different from those in the case we are discussing.) There remain in the deck forty-seven cards. The possible cases are therefore the three-combinations of 47, or $47 \times 46 \times 45$ divided by 3×2. The result is 16,215.

The favorable cases are now to be found. The three cards drawn must contain a pair (not aces), and one other card which is not to be either an ace or of the same denomination as the pair. The forty-seven cards left in the deck consist of two aces, three kings, three queens, three jacks, and four each of the remaining nine denominations. The pair of aces must be excluded from the list of favorable cases, *since their presence would make fours, not two pairs.* The number of favorable cases involving a pair of kings, queens, or jacks is evidently different from the number involving one of the other pairs, say a pair of nines, and we must count them separately. The number of favorable cases involving a pair of kings is 3×42, as there are three distinct pairs that can be made from the three kings. The 42 represents the number of cards remaining in the deck which are not aces or kings (47 minus 2 minus 3). Similarly for queens and jacks, making in all $3 \times 3 \times 42$ favorable cases of the kind. To these we must add the favorable cases involving a pair of rank below the jack. Here there are four cards of each denomination, and the number of pairs of any one, say of nines, is the two-combinations of four, which is six. Thus for each denomination there are 6×41 favorable cases, and as there are nine such denominations, the total is $9 \times 6 \times 41$. The last figure is 41 instead of 42, as before, because the pair has been selected from among four cards instead of three. Gathering these results together, we find that there are 2,592 favorable cases out of the 16,215 possible ones. The probability, then, of the player's securing a final hand containing two pairs is

$$\frac{2,592}{16,215}$$

or approximately 16/100. The chance is 1 in 6.25.

When the player holds an extra card or "kicker" with his pair, drawing only two cards, the chances for two pairs are computed in exactly the same way. Suppose that his original hand consists of a pair of jacks, an ace, a king, and a queen, and that he throws away the last two. The possible cases are $C_2^{47} = $ (47×46) divided by 2, or 1,081. The favorable cases are $(3 \times 42) + (3 \times 2) + (9 \times 6)$; the first term is the number of cases with "aces up," the second the number of those with a pair of kings or queens and a pair of jacks, the last the number of those with a pair of jacks and a lesser pair. The probability of getting two pairs is therefore 186/1,081, or about 17/100. The chance is roughly 1 in 5.8.

So it appears that when the extra card is held there is an additional chance in 100 to fill to two pairs. The odds in the two cases are respectively 5.25 to 1 and 4.9 to 1 against the draw. This advantage is of course more than compensated for in all of the other possibilities of improvement, threes, fours, and fulls, as shown in Table VII on page 126.

We shall illustrate draws to a pair by one further case, the chance of getting three of a kind on a three-card draw. Here the computation of the probability differs slightly from the preceding ones. Again let us imagine that the player holds a pair of aces, discarding three other cards, say a king, a queen, and a jack. The possible cases are, as before, 16,215. The favorable cases are those where the three cards drawn contain one ace and two other cards not forming a pair. There are two aces; so we begin the enumeration by writing $2 \times \dots$. Now of the forty-seven cards to be considered two are aces; there remain forty-five. One of the three cards drawn must be an ace, if the draw is favorable, and if the remaining two are any two of these forty-five cards, the resulting hand will contain three aces, at least. If these cards form a pair the hand will be full, and this result we must exclude from the favorable cases since it is to be computed under the heading of fulls. It is necessary to subtract from the total number of two-combinations of 45 the number of ways in which a pair can be formed from these forty-five cards. There are three kings, three queens, three jacks, and four each of the remaining nine denominations, so that the total number of possible pairs is $(9 \times 6) + (3 \times 3)$, or 63. This we subtract from $C_2^{45} = 990$, which gives 927. Not overlooking the $2 \times \dots$ from above, we obtain for the total number of favorable cases 1,854 and for the probability of threes 1,854/16,215. The chance is about 1 in 9.

The calculation of the chances in the other cases is exactly similar, and the reader who is interested will be able to verify Table VII for himself. The best verification of the correctness of these results is obtained by computing directly the

TABLE VII

Odds against Improving Poker Hands on the Draw

Original Hand	Cards Drawn	Improved Hand	Prob- ability	Odds Against
One pair	3	Two pairs	0.160	5.25–1
	3	Triplets	0.114	7.7 –1
	3	Full house	0.0102	97 –1
	3	Fours	0.0028	359 –1
	3	Any improved hand	0.287	2.48–1
	2	Two pairs	0.172	4.8 –1
	2	Triplets	0.078	11.9 –1
	2	Full house	0.0083	119 –1
	2	Fours	0.0009	1,080 –1
	2	Any improved hand	0.260	2.86–1
Two pairs	1	Full house	0.085	10.8 –1
Triplets	2	Full house	0.061	15.4 –1
	2	Fours	0.043	22.5 –1
	2	Any improved hand	0.104	8.6 –1
	1	Full house	0.064	14.7 –1
	1	Fours	0.021	46 –1
	1	Any improved hand	0.085	10.8 –1
Four straight (1 gap)	1	Straight	0.085	10.8 –1
(open)	1		0.170	4.9 –1
Four flush	1	Flush	0.191	4.2 –1
Three flush	2		0.042	23 –1
Two flush	3		0.0102	97 –1
Four straight flush (1 gap)	1	Straight flush	0.021	46 –1
	1	Any straight or flush	0.256	2.9 –1
Four straight flush (open)	1	Straight flush	0.043	22.5 –1
	1	Any straight or flush	0.319	2.1 –1

probability of *not* improving the hand in question by the assumed draw. If this last is added to the sum of the probabilities of improving the hand in all possible ways, the result should be exactly 1, since it is certain that one of two things will happen; the hand will be improved in one of the possible ways, or it will not be improved. This explains our care, in computing the chances of drawing a certain hand, to exclude higher hands. In this way the various cases are "mutually exclusive," in the sense of Chapter VI, and the above verification becomes possible.

Take the case of drawing three cards to a pair, say a pair of aces. In computing the probability of *not* improving, the favorable cases are those in which the hand contains exactly the original pair after the draw. The number of possible cases is $C_3^{47} = 16,215$. The number of possible three-card draws not containing an ace is $C_3^{45} = 14,190$, and the number of favorable cases is obtained from this last by subtracting the number of ways in which these three cards contain threes or a pair, or $(9 \times 4) + (3 \times 1) + (9 \times 6 \times 41) + (3 \times 3 \times 42)$, 2,631 in all. Subtracting this from 14,190 gives 11,559, so that the probability of not improving a pair on a three-card draw is

$$\frac{11,559}{16,215},$$

or .713. From the table of chances it is seen that the probability of improving a pair on a three-card draw is .287. The sum of the two is 1, verifying the correctness of the chances as computed.

In finding the chance of making two pairs on a three-card draw to a pair, we assumed that the discard did not contain a pair. As mentioned at the time, it sometimes happens that the player prefers to throw away the smaller of two pairs in order to increase his chance of ending up with a hand able to beat any two pairs. It is clear from the manner in which the favorable cases were enumerated that this fact changes the probability to some extent. As a matter of fact, the change is entirely too small to have the least practical bearing, but it may be mentioned under the head of curiosities that the player's chances of bettering his hand are actually slightly improved when he has discarded a pair instead of three cards not containing a pair. In the former case the probability of improvement is 4,695/16,215, in the latter 4,656/16,215.

The problem of finding the probability of holding a given hand on the deal is one that offers no special difficulty. It is merely a matter of enumerating, in each case, the favorable five-card hands. The total possible cases consist evidently of all possible poker hands. Their number is C_5^{52} as explained on page 121; the number of possible different poker hands is 2,598,960.

To illustrate these calculations, let us find the probability of holding a four flush, one that is omitted from the tables of poker chances that I have seen. Suppose first that the four flush is to be in spades. Four of the five cards must be selected from the thirteen spades and the remaining one can be any one of the thirty-nine cards of the three other suits. There are $C_4^{13} \times 39$ favorable cases for a spade four flush. For a four flush in any suit there are $4 \times C_4^{13} \times 39$ favorable cases out of the 2,598,960 possible ones, so that the probability is

$$\frac{111,540}{2,598,960},$$

or about 43/1,000, 1 chance in 23. This includes the probability of holding a four-straight flush (open or not), which last is represented by a very small number, in fact

$$\frac{123}{1,000,000}.$$

A player may expect, then, to hold a four flush once in about twenty-three hands, on the average. We saw above that his chance of filling a four flush is 1 in 5⅔, so that he may expect to fill a four flush once in one hundred and twenty hands on the average, provided that he draws to every one. If we wish to know how often he may expect to hold a flush, we must take account of his chance for a pat flush. From Table VIII it is found that the chance of a flush on the deal is 1 in 509. Therefore out of three thousand hands he should on the average hold six pat flushes and fill twenty-five four flushes, or thirty-one flushes in all. He may expect on the average one flush in ninety-seven hands, always provided that he never refuses to

TABLE VIII

ODDS AGAINST HOLDING VARIOUS POKER HANDS ON THE DEAL

Hand	Probability	Odds Against
One pair	0.423	1.4–1
Two pairs	0.048	20 –1
Triplets	0.021	46 –1
Straight	0.0039	254 –1
Flush	0.0020	508 –1
Full house	0.0014	693 –1
Fours	0.00024	4,164 –1
Straight flush	0.000014	72,192 –1
Royal flush	0.0000015	649,739 –1
Four straight	0.167	4.9–1
Four flush	0.043	22 –1
Four straight (open)	0.035	27 –1
(middle)	0.123	7.1–1
(end)	0.0087	114 –1

draw to his four flushes. Combining the two tables of chances in this same way, it is easy to find out how frequently a given player may expect to hold the various hands, but in cases where the hand may be obtained by different draws (holding a "kicker" etc.), the proportion of each must be taken into account.

Even in the case of the four flush, where there is only one possible draw, the habits of the player must be considered; for no player draws to every four flush. In fact, Hoyle remarks, "The best players never draw to four-card flushes except when they have the age, and the ante has not been raised." (To "have the age" means to sit at the immediate left of the dealer. In the variety of poker to which this passage refers the "age" puts up a "blind" before the hand is dealt.)

The reason for this statement of Hoyle is that good players invariably take account of the amount of money already in the pot and the amount they are required to add to it, as well as the abstract poker chances. When a player has the age, and the ante has not been raised, it costs him no more to play the hand than not to play it. He therefore has everything to gain and nothing to lose and will surely play the hand through. The general question of under what conditions players should stay in the hand is one of the more interesting probability problems of poker and will be considered at some length in the next chapter.

When the player finds himself in a position where it costs him no additional chips to play the hand, he will draw cards, even though his hand contains not so much as a small pair. The chances, if he draws five cards, are exactly the same as his initial chances on the deal and are therefore given in Table VIII on page 129. This is correct, at least, provided that he has received no significant information during the hand. If he knows from their discards that a certain number of the players were dealt pairs, or some other significant hand, his chances will be slightly different from those given in the table, but the differences will be small. Instead of drawing five cards, the player may draw four cards to an ace or a king. It is not difficult to compute his chances of improving his hand in this case, but we shall drop the matter at this point.

The probabilities centering in the deal itself illustrate once again the fact that the probability of a given event depends on the cards *seen* or known by the player whose chances we are computing, and not at all on which cards are already in other players' hands or in the discard heap. In particular it has nothing to do with the manner in which the cards are dealt (one at a time, five at a time, etc.) , and is the same whether your hand is the first or the last dealt, always assuming that the cards are thoroughly shuffled, and that the game is honest.

Suppose, for instance, that there are two players and that five cards are dealt to the first player before the second gets his hand. The chance that the second player (or the first) has a pair of *aces* is about 1 in 31 (1/13 of 1/2.4). Suppose the second player admits that the chance that the first hold a pair of aces is 1 in 31 but in his own case argues as follows: "The first hand may hold one ace, or two, three, or even four aces. In each of these cases the chance of my hand's holding exactly a pair of aces is different, and if the other hand contains no aces, clearly my chances are improved. So I am not satisfied with the 'blanket' figure of 1 chance in 31; I should like to see the analysis for each of these possibilities separately."

Let us attempt to satisfy the player's skepticism, which he does very well to bring out into the open. We shall ask him in return to meet us halfway by permitting us to simplify the situation as much as possible without destroying the significance of the problem, as we have a perhaps pardonable aversion to unnecessary complications. Instead of using a full pack of fifty-two cards we shall consider one containing only twelve, three each of aces, kings, queens, and jacks, and instead of a hand of five cards we shall consider hands of three cards only. The situation is simplified but not changed in any material respect. The possible hands are reduced from 2,598,960 to C_3^{12}, or 220. We must know first of all the probabilities that the *first* hand will hold one ace, two aces, three aces, or no aces. These are easily found to be respectively 108/220, 27/220, 1/220, and 84/220. Note that their sum is 1. Now consider the second hand. If the first contains either two or three aces, the chance that the second contains a pair of aces is evidently

0; it is impossible. If the first hand contains one ace, there are two left in the deck, as well as seven other cards. The favorable cases for the second player are the three-card hands containing the pair of aces and one of the other cards; there are 7. The possible cases are the three-combinations of nine, which number 84. The probability in this case is $\frac{7}{84}$, or $\frac{1}{12}$. There remains the case where the first hand contains no aces. The possible second hands are 84, as before. Of the nine cards left in the deck, three are aces and six not. The favorable cases are any pair of the three aces, three in number, combined with any one of the six other cards, making a total of 18. The probability in this case is $\frac{18}{84}$, or $\frac{3}{14}$.

To find the chance that the second player will hold a pair of aces, it remains to combine these results in the proper way. Consider the case where the first player holds exactly one ace. The chance of this is 108/220, or 27/55. On the assumption that this has already taken place we have just seen that the chance that the second player holds a pair of aces is $\frac{1}{12}$. Therefore the probability that these events happen in succession is $\frac{27}{55} \times \frac{1}{12}$, which is equal to 9/220. In the same way, when the first player holds no aces the probability of success is $84/220 \times \frac{3}{14}$, which equals 18/220. When the first player holds any other number of aces we have seen that success is impossible. Therefore the chance that the second player holds exactly one pair is 9/220 plus 18/220, which is equal to 27/220. The probability that the first player holds exactly one pair was also found to be 27/220. We have shown directly that whether or not a previous hand has been dealt, the probability of holding a pair of aces (in a three-card hand from a twelve-card deck) is the same, a result which should settle the doubts of our skeptical player.

The probability of a hand that contains no pair or better, but may contain a four straight or a four flush, is almost exactly $\frac{1}{2}$; the odds are even.

Tables VII and VIII give the chances of the significant

hands, respectively, for the draw and for the deal. Table VIII tells us, for example, that there is about one chance in twenty that the man on your left, or any other player, was dealt a hand with two pairs. This means that in an eight-handed game you would expect some one of your opponents to be dealt two pairs about once in three deals. The table gives a similar result for any other type of hand. For example, in a six-handed game you will expect two of the other players, on the average, to hold one pair on the deal; you will expect one of the five to hold a four straight. We have not yet learned how to combine these probabilities, how to find, for example, the probability that two of the players will hold pairs *and* one player a four straight. This problem will be discussed in the following chapter.

Table VII tells us many things about the chances on the draw that have considerable importance in playing the hands. We shall illustrate this remark by one example; the interested reader will discover others for himself. Suppose that your hand contains one pair, an ace, and two indifferent cards. It is your turn to discard. You must decide whether to draw three cards to the pair or two cards to the pair and a kicker, the ace. As pointed out on page 122, good players make a point of varying their play by occasionally holding the extra card. The ideal occasion for making this play is when it gives the best chance for winning the hand. Now Table VII tells us that the chance of two pairs after the draw is actually larger if you hold the kicker. This chance is about 1 in 6. If the kicker is an ace, the chance that you will get two pairs with aces high is two thirds of this, or 1 in 9. So if you find yourself with such a hand and if the other players are making one-card draws, it is shrewd play to hold the ace kicker. For a little reflection will show that under these circumstances you have almost as good a chance of winning the hand with two pairs, aces up, as you do with threes, and you have more chance of getting the former if you hold the ace kicker and draw two cards.

This play becomes almost a must if your original pair is of low rank. In fact, unless the number of players in the game is small, it is unwise to draw three cards to a low pair, if doing so involves a contribution to the pot.

CHAPTER *XI* ·

Poker Chances and Strategy

In the last chapter we applied some of the ideas of probability theory to the game of poker and found, in particular, the odds against improving various hands on the draw and those against a certain player's holding a given hand on the deal. These are the fundamental chances on which all further reasoning about the game must be based. The questions that we come to now are perhaps the most interesting of all from the standpoint of the theory of chance and, incidentally, those with the most direct bearing on poker strategy.

The first examples of these problems have to do with the chance that several players hold a certain type of hand—pairs, for example—simultaneously. The second are concerned with the monetary risks that a player is justified in taking, and it will be shown that the use of the idea of expectation, as given in Chapter VII, leads to a simple rule, which is often at variance with the widely accepted principles.

Our first problems are of the general nature of this one: With three players in the game what are the odds that at least one of them will hold a pair of tens or better before the draw? The next sort of question is even closer to the concrete problems of the game: In a seven-handed game the player at the dealer's left holds a pair of jacks after the deal. What is the chance that at least one of the remaining six players is holding a stronger hand? We base our answers to these questions on the table of chances on the deal given on page 129. Here we find, for example, that the odds against the player's being dealt a pair are 1.4 to 1. His practical interest in this figure is primarily as an aid to finding the chance that no other hand will be higher than his, so that he can decide how much the privilege of drawing cards is worth to him.

Before attacking these problems it is as well to point out that they involve difficulties that were not present in the relatively simple poker chances that we have computed so far. In previous cases it was possible to enumerate directly the possible and the favorable cases and to determine the probability in question by dividing the latter by the former, according to the elementary definition of the word *probability*. In the present problems the use of this direct method would require classifying the favorable cases in hundreds of different ways; the labor involved would be prodigious and it is doubtful whether anyone would be willing to make the effort, no matter how little else he had to do. We have already encountered one example of a problem which could not be handled by the direct approach, that of the ruin of a gambler discussed in Chapter VIII. In that instance it was not possible to list the cases, because they are infinite in number—the list could never be finished. In the present case the number of combinations is limited (though large), but the analysis is too complicated for us; theoretically the list could be finished, but it would require an almost incredible amount of time and effort.

We shall have to find a way around this difficulty, unless we are to omit the study of many of the most interesting examples of the theory of chance that come from games. And in order to make clear what our problem is, and to help us in avoiding the danger of falling into errors of reasoning, it may be best to stop a moment to consider the basic principles that we used in the preceding chapter in computing poker chances. We can state them as follows: (a) The probability that a given player holds a certain hand, such as exactly a pair, does not depend on the manner of dealing the cards, whether one at a time, which is the correct method for poker, or two at a time, or five at a time. (*b*) This probability does not depend on the number of players in the game, or on any player's position at the table. (*c*) The probability of an event (on the deal or draw) relative to a given player depends on the number and position of those cards which he has seen or otherwise knows, and not at all on the positions of the other cards, whether they are in the pack, in other hands, or in the discard. The first two principles are in reality different ways of stating the same thing.

As each player picks up his hand, then, his individual probability of holding a pair (and no more), for instance, is the same as that of each other player; as a matter of fact it is 423/1,000 or, what is equivalent, the odds are 1.4 to 1 against. What is the chance, then, that at least one of the players holds a pair? Let us first consider a game with only two players, where the situation is simple enough to be handled by the direct method of counting cases.

For each player separately the chance of holding a pair is 0.4226. Therefore the chance that he will not hold a pair is 1 minus 0.4226 or 0.5774. If we were not on our guard we might be inclined to argue as follows: The probability that at least one of the players holds a pair can be obtained by first finding the probability that neither holds a pair and subtracting the result from 1, since it is certain that one of the possible cases takes place. The probability that neither holds a pair is 0.5774 × 0.5774, which gives 0.3334, or almost exactly ⅓. Subtracting this from 1, we get for the required probability ⅔.

This reasoning is incorrect. It assumes, in fact, that *after* we know that one of the hands does not contain a pair, the probability that the second does not contain a pair is the same as though we knew nothing about the first hand. This is a violation of the third of the principles just enumerated. To be strictly accurate, account must be taken of the change in the probability for the second hand due to our further knowledge of the first hand. In other words, the two events are not *independent* of each other, in the sense of Chapter VI.

The reader may be troubled at this point by the following objection: As the cards are, in reality, dealt one at a time, there is no first player in any real sense; yet in the above argument we spoke of the first hand as though it had been dealt before the second hand. It is true that there is no real distinction from the present point of view between the two players, and that anything that we discover about the chances of one of them applies equally to the other. It is, however, simpler to imagine that one of the hands is dealt out before the other, and according to the first of the three foregoing principles the manner of dealing has no effect on the probabilities; we are at liberty to consider one of the hands as first if we wish to do so. If this plan is not adopted, it is necessary to enumerate the possible pairs of five-card hands and to pick out those favorable to the event under consideration, a hopeless undertaking when we come to games involving a number of players.

We have used an incorrect argument to compute the probability that at least one of two players holds a pair, obtaining as a result $\frac{2}{3}$. Are there any conditions under which this argument would become correct? When we were discussing the chances in throwing two dice we made use of the identical argument in question here. This reasoning was correct there because the throw of the second die is unaffected by that of the first. The throws are independent. Similarly in poker we can imagine a change in the rules that would make the situation entirely analogous. Suppose that five cards are first dealt to player number 1, who records them on a sheet of paper and returns them to the dealer. The latter now shuffles the deck thoroughly and deals five cards to player number 2 (we are still confining the game to two hands). Now it is clear that under these conditions the chance that the "second" player holds a pair is the same no matter what the other hand contained, for every condition of the deal is the same for both hands. When the cards are not put back, however, we may think of the second hand as dealt from a pack of only forty-seven cards, and the number of possible pairs in it depends on which five cards were dealt out first. And if we have any knowledge as to these cards, the probability of a pair in the second hand is changed.

The foregoing argument which, as we now see, amounts to assuming that each hand (except the last) is returned to the deck before the next hand is dealt, has all the virtues except correctness. Suppose that we set ourselves the problem of discovering how to allow for the error in those cases in which we are interested, or to prove that the error is small enough to be neglected, so as to be able to use this very convenient tool which does most of the work without effort on our part. This investigation is entirely outside the scope of this book, but its results are fortunately very simple, and we shall make full use of them.

It turns out that in the very cases in which we are most interested, the error introduced by the use of the incorrect argument is so small that it is of no practical importance. We shall therefore use this method, which we shall call, to distinguish it, the *replacement* method. Before doing so we shall give some examples where both the correct and the replacement method can be carried through, and compare the probabilities obtained. Furthermore, these examples will furnish at least a clue as to why the latter (incorrect) method gives results so close to those of the former. Some readers may prefer to skip this section, which is not essential for what follows.

We consider first a two-handed game and ask the probability that at least one of the players holds a pair. This is the problem that we solved by the replacement method, getting as the probability almost exactly $\frac{2}{3}$. We are now interested in solving it by the correct method, so as to compare the two results. We already know the probability for the *first* hand (the cards are dealt five at a time) to hold exactly a pair; it is 0.4226. It will not be necessary to include all of the details of the solution; the reader who enjoys problems can supply them for himself. It can be shown, and we shall take advantage of the short cut, that the probability that at least one of the players holds a pair is equal to twice the probability for the first player to hold a pair (0.4226), minus this same probability multiplied by the probability that the second player holds a pair, computed on the assumption that the first hand already contains a pair. To have our result we need only the last-mentioned probability. Since the first hand contains a pair, say of aces, and three cards not making a pair or threes, there remain in the deck forty-seven cards, of which two are aces; the remaining cards consist of three denominations of three cards each and nine of four cards each. The probability of a pair is easily found to be

$$\frac{650,163}{1,533,939},$$

or 0.4239. (Compare this with the calculations of the preceding chapter.) The probability that at least one of the players has a pair is therefore $(2 \times 0.4226) - (0.4226 \times 0.4239)$, which gives a result of 0.6661. This is to be compared with the result computed by the replacement method, which is 0.6666. The replacement method overstates the probability by 0.0005, an error of about $\frac{1}{13}$ of 1 per cent.

It is to be noted particularly that the change in the probability due to the knowledge that the first hand contains a pair is small, in fact $+ 0.0013$. It is in part due to this fact that the replacement method gave such accurate results. There is also another reason. If the first player holds a hand with no two cards of the same denomination, the probability that the second hand contains a pair comes out to be 0.4207, which is less than the probability for the first hand, while in the case where the first hand held a pair we got 0.4239, which is greater. The average of these two values is 0.4223, which is very nearly the probability of a pair for either hand, when no information about the other is given, namely 0.4226.

This fact indicates that there is a partial compensation for the error introduced by the replacement method. Something of the sort takes place in every situation of the kind; for the compensation is brought about by the fact, noted above, that the two players are on exactly the same footing; we can compute the total probability for the second player to hold a pair according to the various possibilities for the first hand, one at a time, and upon combining them properly, as was done on page 131, we must obtain the result 0.4226.

A rather curious fact is brought out by these calculations, one which is perhaps worth a short digression. If one were asked whether a pair is more likely in the second hand when there is a pair in the first or when there is no pair in the first, he would be almost sure to indicate the second alternative. He might feel that if the first hand contains a pair, "one of the possible pairs has been used, reducing the possible number for the second hand." Yet we have just seen that the opposite is the fact; with a pair in the first hand, the probability of a pair in the second hand is *increased*. Furthermore, with one exception, the better the first hand, the greater the probability of a pair in the second (straights and flushes are to be classed in this connection with hands of no value, for obvious reasons). The application of this to the game is as follows: When you find that you hold a strong hand after the deal, you know that the chance of strong hands against you is very slightly increased. In connection with this point we have computed the figures in Table IX.

It was shown above that for a two-handed game the probability that at least one of the players will hold a pair is very accurately given by what was called the replacement method. For games with more than two players the complications of the direct method multiply rapidly, and to illustrate the accuracy of the method in these cases we turn to other situations. It might be thought that if attention is restricted to pairs of but one denomination, say aces, the approximate method would not give good results, and it is true that the error is larger in this case. Even here, however, the replacement method meets the needs of practical poker strategy.

TABLE IX

PROBABILITY OF A PAIR FOR OTHER PLAYERS WHEN YOUR HAND
AFTER THE DEAL IS AS INDICATED:

Your Hand	*Probability of Pair for Other Hands*	*Odds Against*
No pair	0.421	1.38 to 1
One pair	0.424	1.36 to 1
Two pairs	0.427	1.34 to 1
Threes	0.431	1.32 to 1
Full house	0.427	1.34 to 1
Fours	0.444	1.25 to 1

Consider first a very simple case of hands of one card only. A favorable event will mean holding an ace. Cards are dealt to the several players; each has 1 chance in 13 of holding an ace. If we know that the first player has an ace, the chance for the other players becomes $\frac{3}{51}$, which is less than $\frac{1}{13}$; on the other hand, if we know that the first player has not an ace, the chance for the other players becomes $\frac{4}{51}$, which is greater than $\frac{1}{13}$. We are interested, as before, in finding the probability that at least one of the players holds an ace. The following table contains the comparison, for various numbers of players, as indicated, of the exact probability and that computed by the replacement method:

We return again to the game of poker and consider briefly

one more problem of the same sort, selected because it admits of direct solution. In a four-handed game we wish to find the probability that at least one hand is dealt containing a pair of aces. In this instance we shall make an exception to our usual interpretation of the preceding statement and agree to count as favorable every hand containing a pair of aces, regardless

TABLE X

PROBABILITY OF AN ACE IN AT LEAST ONE OF INDICATED NUMBER
OF ONE-CARD HANDS

Number of Hands	Correct Probability	Approximate Probability	Error Per Cent
2	0.149	0.148	0.9
3	0.217	0.214	1.8
4	0.281	0.274	2.6
5	0.341	0.330	3.3
6	0.397	0.381	4.0
7	0.450	0.429	4.6
8	0.499	0.473	5.2

of the other three cards. Thus in the present case three aces, or an ace full and so on, will be included among the favorable hands. The correct probability, computed directly, comes out to be 0.154; if we employ the replacement method the result is 0.150. The error is 2.7 per cent.

It is to be remembered that each of the last two illustrations, those concerned with cards of one denomination only, is among the less favorable situations for the application of this method. In the usual applications we shall expect much closer agreement.

Now that we have acquired a simple method that permits us, at least with certain precautions, to attack problems having to do with several players, let us apply it to some of the situations that come up in actual play. First of all we turn our attention to the game of jack pots. What is the probability that the pot will be opened? From Table VIII on page 129 we find that the chance that a player will hold one pair or better on the deal is 0.499, or almost exactly $\frac{1}{2}$. The chance that any given player will be able to open the pot is found by subtracting from $\frac{1}{2}$ the probability that his hand will contain exactly one pair of tens or less. The probability that his hand will contain one pair is 0.423, and the probability that the pair will be tens or less is $\frac{9}{13}$ of this amount, or 0.292. His chance of opening the jack pot is 0.207, or about $\frac{1}{5}$.

Using the replacement method, we proceed as follows: The chance that a given player will not be able to open the pot is $\frac{4}{5}$; therefore the chance that two players will not be able to open is $\frac{16}{25}$. The chance that one of the two players will open is therefore $1 - \frac{16}{25}$ or $\frac{9}{25}$, or 0.36. Continuing in this way we get the following results: For three players the chance is 0.49; for four players it is 0.59, for five 0.67, for six 0.74, for seven 0.79, and for eight 0.83. With five players the pot will not be opened about one deal in three on the average, with eight about one deal in six. With less than six players it is a waste of time to play jack pots.

In playing jack pots the first problem that confronts the player is under what conditions is it wise to open the pot, when his hand is only just strong enough to do so. Suppose that the game is a seven-handed one, and that the player sits immediately at the dealer's left, "under the guns," as it is sometimes called. He finds that his hand contains a pair of jacks and three odd cards. What is the chance that one, at least, of the other six players holds a pair at least equal to his? The first step is to find the chance that any *one* of these players has of holding such a hand, and to do this it is essential to specify the denominations of the three odd cards in the first player's hand; at least we need to know which of them are aces, kings, or queens. This divides the problem into four separate ones, which can be indicated by writing the player's hands, omitting suits, which are immaterial. For instance, the first case will be hands like J J 10 9 8, or J J 8 5 4, in which none of the odd cards are aces, kings, or queens. The second case will be hands like J J A 10 9, or J J A 3 2, and will include all hands like J J K 10 9, or J J Q 10 9. With this understanding the results are as follows: Case I. Typical hand J J 10 9 8. Probability for other players 0.125, or ⅛. Case II. Typical hand J J A 10 9. Probability for other players 0.107, or about ⅑. Case III. Typical hand J J A K 10. Probability for other players 0.089, or about $\frac{1}{11}$. Case IV. Typical hand J J A K Q. Probability for other players 0.071, or about $\frac{1}{14}$.

The first point to be noticed in this listing is the great importance of the three odd cards in the first player's hand. If these cards are AKQ, the probability for each of the other players is diminished by 43 per cent, as compared to the case when all are below the jack in rank (Case I). The player should not fail to take account of this fact in playing jack pots. Similarly, although the above figures apply only to hands containing exactly one pair, it is clear that the same is true, though to a less marked extent, with respect to stronger hands (always omitting straights and flushes), and that it is easy to compute the exact probabilities for all of these cases.

We are now in a position to determine the chance that *at least one* of the other six players holds a pair of jacks to aces, inclusive. First for Case I: The chance that any particular player will not hold one of these pairs is, by the above, 1 — .125, or ⅞. Using the replacement method, the chance that all six hands will not contain one of these pairs is ⅞ raised to the sixth power, which is

$$\frac{117,649}{262,144},$$

or about 45/100. The probability that at least one such pair will be held by the six players is therefore 1 — 45/100, or 55/100, a little better than even. The corresponding results for the other cases are as follows: Case II, 51/100; Case III, 44/100; Case IV, 36/100.

If the question is slightly changed, and we ask the probability that at least one of the other players will hold a pair *better* than jacks, these results become as follows: Case I, 53/100; Case II, 47/100; Case III, 40/100; Case IV, 32/100. What have we learned concerning the correct play of the man at the dealer's left? *If pairs higher than jacks were the only hands to be considered,* the conclusions would be as follows: When the player can open for an amount not greater than the amount already on the table, he can do so with favorable prospects, if his hand is one of those we have called III and IV. Case II has been omitted; it is near the dividing line, and the effect of four straights and four flushes has been left out of account, hands to which players incline to draw much too frequently. When the player can open for appreciably less than the amount in the center, he is justified in doing so in any of these cases. The amount on the table is of the greatest importance in all problems of poker strategy, and something more will be said later concerning the correct manner of taking account of this factor.

When account is taken of hands higher than a pair of aces, these figures are considerably modified. We can do this accurately enough for practical requirements by neglecting the very small change in the probability of these hands, due to our knowledge of the first player's hand. Table VIII on page 129 shows that the probability of a hand better than a pair of aces is 0.075. The results now become: Case I, 72/100; Case II, 68/100; Case III, 64/100; Case IV, 59/100. If the player has to pay an amount equal to that already in the pot, he should not open. We can also assert that in any one of the cases he should not pay as much to open as the amount that is already in the pot, divided by the odds against the player. If p is his probability of winning the hand, the odds against him are $(1 — p)/p$. For example, if the player holds a hand under Case II, which is one like J J A 2 3, and the amount in the pot is $2, he should not pay more than 95 cents to open, while if his hand comes under Case I, he should not pay more than 80 cents. For the rest, the maximum is: Case III, $1.10, Case IV, $1.40.

The practical result is as follows: The player should not make a practice of entering pots when the conditions just laid down cannot be satisfied; for if he does, he is certain to lose on such pots in the long run. The question of whether, when these conditions are satisfied, he should or should not open the pot, is of very much less importance, due to the fact that with six players coming after him, the chance that the pot will not be opened is not large. Some players make it an almost invariable rule not to open the pot on a pair of jacks, regardless of the balance of their hand, when seated at the dealer's left. This is being unnecessarily cautious, as we have seen before. Another fairly common practice is to pass a stronger hand, such as two pairs or threes, when in this position, anticipating that the pot will be opened by a weaker hand and that the ensuing one-card draw will be taken to indicate a four straight or four flush. This is sometimes an effective play, especially against weak players, but is scarcely so against the player who is on the lookout for just such maneuvers. Nevertheless, like many similar practices, it becomes good play when occasionally indulged in for the sake of variety, for it is very bad poker to play a "mechanical" game which opponents soon learn to read· almost as well as their own cards.

It is interesting to compare the figures we have cited with those obtained when the player holds a pair of queens instead of jacks, and the latter are very easily found from the former. There are now but three cases, represented by hands of the following types: Case I, Q Q 10 9 8; Case II, Q Q A 10 9; Case III, Q Q A K 10. The probabilities that at least one of the remaining six players will hold a hand stronger than a pair of Queens, in the respective cases, come out to be: Case I, 63/100; Case II, 58/100; Case III, 53/100. It is clear that in the case of a pair of queens also, except possibly when the other three cards are such that the hand comes under Case III, the player should not open if he has to put up an amount equal to that already in the center. The rule for what fraction of this amount he is justified in putting up is the same as before.

When the player at the dealer's left has a pair of kings, there are two cases: I, of type K K 10 9 8, II, of type K K A 10 9. The probabilities of at least one hand stronger than a pair of kings come out: Case I, 52/100; Case II, 45/100. When the amount required to open is equal to that on the table it is surely sound to do so in Case II, and a borderline case in the other. When the player holds a pair of aces the probability of one or more better hands against him is 37/100. He has a decided advantage in opening the pot.

All of these results apply to the probabilities of your hand in a seven-handed game *after* you have examined it and found one of the pairs mentioned, regardless of your position at the table, provided that no player has as yet either passed or opened. We considered the player as seated at the dealer's left because we were talking of jack pots and the rules for opening them. As a matter of fact we have answered the following question: In a seven-handed game what hand must a player hold before the draw in order that the odds be even that he has the best hand out? The answer is that there are two types of hands which come very close to fulfilling this condition. They are the hands of the following types: Q Q A K 10 (Case III under queens), and K K 10 9 8 (Case I under kings).

If the player in question does not sit next to the dealer on the left, the problem of his opening a jack pot does not come up unless all of the players preceding him have passed. As the fact that these players have passed is, to some extent, information regarding their hands, we should expect it to change the probabilities that the player's hand is the best one out. This change is small, however, and it is not worth while to take account of it, especially in view of the fact that there is no guarantee in an actual game that one or more of the preceding players has not passed a hand with a pair of jacks or better. If the player sits in the third chair to the right of the dealer, for example, there are four players to follow him, including the dealer, and we may consider the problem the same as that for the player at the dealer's left in a five-handed game. To illustrate how the chances vary with the number of players, we include the following table:

TABLE XI

PROBABILITY THAT INDICATED HAND IS *Not* THE STRONGEST OUT
FOR DIFFERENT NUMBERS OF PLAYERS

Hand	#	Prob.	#	Prob.	#	Prob.	#	Prob.	#	Prob.
J J 10 9 8	7	0.72	6	0.66	5	0.58	4	0.48	3	0.35
J J A 10 9	7	0.68	6	0.62	5	0.54	4	0.44	3	0.32
J J A K 10	7	0.64	6	0.57	5	0.49	4	0.40	3	0.29
J J A K Q	7	0.59	6	0.53	5	0.45	4	0.36	3	0.26
Q Q J 10 9	7	0.63	6	0.56	5	0.49	4	0.39	3	0.28
Q Q A J 10	7	0.58	6	0.52	5	0.44	4	0.36	3	0.25
Q Q A K J	7	0.53	6	0.47	5	0.39	4	0.31	3	0.22
K K Q J 10	7	0.52	6	0.45	5	0.38	4	0.30	3	0.22
K K A Q J	7	0.45	6	0.40	5	0.33	4	0.26	3	0.18
A A K Q J	7	0.37	6	0.32	5	0.27	4	0.21	3	0.14

This table contains several very interesting things, and it is worth while to take the trouble to learn to read it correctly. Each probability, corresponding to a definite type of hand, classified according to the rank of the pair and those of the three odd cards, if of higher rank than the pair, and to a definite number of players, indicates the chance that one or more of the other players holds a stronger hand. If any one of these probabilities is subtracted from 1, the result is the chance that the player's hand is the strongest out. The table is therefore in effect a classification, in order of potential strength, of the hands that are included, the smaller the probability corresponding to a given hand, the stronger the hand. In this way it is seen that any one of the hands like J J A K Q, of which there are 384, is in this sense stronger than those of the type Q Q 4 3 2, of which there are 2,880. For a hand to be stronger "in this sense" means precisely this: There is less chance that one or more of the other players is holding a better hand before the draw.

Similarly, hands of the types Q Q A K 2 and K K 4 3 2 are of almost the same strength. If you draw a line across the table intersecting each column where it reads nearest to 0.50, it will run from the lower part of the left column to the upper right, missing the last two columns. This line indicates the hands, for different numbers of players, that have an even chance of being the best out. From its position it can be seen without any calculation that in three-handed and four-handed games it is hands containing a pair of tens that have an even chance of being the best. In a four-handed game it is one of the type 10 10 A K Q, and in a three-handed game one like 10 10 A 9 8.

Presumably the choice of jacks as openers in jack pots was an attempt to give the player opening the pot at least an even chance of going into the draw with the best hand, and our figures show that this is actually the case for a five-handed game. For a six-handed game a pair of queens would be the correct minimum hand, and for a seven-handed one a pair of kings. An experienced poker player whom I consulted on this matter told me that his impression from watching the cards is that a pair of kings has even chances of being the best in an eight-handed game, a very correct observation. He was not aware, on the other hand, of the full importance of the rank of the three odd cards in the hand, although, like all good poker players, taking account of it to some extent.

While on the subject, there is one other figure that might

be of interest, and that is the chance that in a seven-handed game where your hand contains a pair of aces, some other player holds a pair of aces as well. This comes out to be about 1 chance in 23. Any other pair could be substituted for aces without changing this figure. All of the above results tend to show the strength of a pair of aces before the draw; a hand that you will hold, on the average, once in about thirty-one deals.

In Hoyle we find no figures on the matter of the relative value of hands before the draw. There is only the following statement: "An average go-in hand is a hand which will win its proportion of the pools, according to the number playing, taking all improvements and opposition into account. This can be demonstrated to be a pair of tens." What is meant by an "average go-in hand" is what we have referred to as a hand whose chances of being the best out before the draw are even. We have seen above that this hand is a pair of tens only in games with either three or four players, so that the statement in Hoyle becomes correct only with this essential supplement.

One of the most important features in poker strategy concerns the amount of money that a player is justified in paying for the privilege of drawing cards, or to open the pot. The clearest approach to these questions is through the use of the idea of *expectation*. As explained in Chapter VII, a player's expectation is the chance that he will win what is in the pot, including what he himself has contributed, multiplied by the amount of the pot. Suppose that at a given moment the player is faced with the decision as to whether he should or should not pay $1 to enter a pot which already contains $4. His decision should not be affected, and this is a point of the very greatest importance, by what fraction of this amount of $4 he himself put in. That is "water over the dam," as the saying is. If he has already paid more than his hand warranted, or if, on

THE SCIENCE OF CHANCE

the contrary, other players have put up the entire amount, his problem is exactly the same: Are his chances of winning the pot better or worse than 1 in 5?

If they are better, he can count on winning in the long run in this situation; if they are worse, he can count equally on losing. If they are exactly 1 in 5, it is an even bet, like one on heads or tails in tossing a coin. What makes the game interesting to the player who enjoys it is in part the fluctuations that distinguish the "short run," as we saw in connection with the game of heads or tails. It is always possible to win several successive pots with the chances against you, or to lose several in which the chances are strongly in your favor. The wise player does not allow the former fact to tempt him into bets when he knows that the odds are in favor of his adversary, nor the latter to give up bets in his own favor.

The average poker player seems to have an almost compelling urge to stay in a pot to which he has already been a large contributor, apparently unable to realize that putting in more chips, when his expectation is less than the value of these chips, does not in any way protect his previous contributions, but is merely making a bad bet that is entirely independent of what he has done up to that time. The same man would indignantly refuse to give odds of 7 to 1 on throwing a 6 with a die, yet he is in effect doing a like thing when he puts up $1 against an expectation of 50 cents.

The point to be kept in mind is this: With each addition to your information regarding other hands during the course of the play, the probability that your hand will win changes. So does the amount in the center. Your expectation varies due to both these factors, and so must your policy in playing the hand.

In actual play many of the most important elements that enable the player to decide what his course of action should be are judgments of what his adversaries are doing, based on their bets, their draws, their habits and characters as known to him, and other things The point I wish to make is, first, that these judgments, in the form in which they are usable, are all estimates of the chance that one thing or another will happen, or has already happened, and second, that these judgments can and should take the form, "The chance of a certain event is 1 in so many." Instead of a cautious vagueness the poker player should cultivate a crisp conciseness; instead of thinking, "I feel quite sure that Mr. A's one-card draw is a bluff," he would do better to express it, "My judgment is that the chances that Mr. A is bluffing are 4 out of 5." It is not important that he get the probability exactly right, a very hard thing to do; what is important is that with a definite numerical estimate, even if it is a little wrong, he can carry his "numerical thinking" through and in many cases arrive at a definite conclusion as to the required action, which would not be changed by a slight change in his estimate that Mr. A is bluffing. It is with a similar thought in mind that Hoyle advises the player to avoid calling a hand. If you consider your hand better, he says, raise the bet; if not, drop out. We shall shortly examine the soundness of this advice.

As an illustration of the advantage of numerical thinking in poker, let us take the case of Mr. A, who is suspected of bluffing, and apply it to a game. A sits at B's right and passes the jack pot which B opens with three aces. A raises, B stays, and the other players drop out. A asks for one card, B two, and the latter bets one white chip. Let us assume that at this point there is $2 in the center of the table. A now raises B's bet $5. B, from what he knows of A's game and from the manner in which the latter has played the hand, makes the definite judgment that the probability that A is bluffing is $2/5$, or that the probability that he is not bluffing is $3/5$. B's draw does not improve his hand. How should B play the hand? To make the situation simpler we assume that B, disregarding the advice of Hoyle, will choose between calling and dropping out. Is it sound play for him to pay $5 to see A's hand?

B knows that *if A is not bluffing*, his own three aces will be beaten, for A was unable to open the pot, and his subsequent play therefore shows that he has filled a straight or a flush, still assuming that he is not bluffing. We must find B's expectation. The probability that B will win is the probability that A is bluffing, which B has decided is $\frac{2}{5}$. If B calls there will be $12 in the pot, so that B's expectation is $\frac{2}{5} \times 12$, or $4.80. If he calls he is paying $5 for $4.80 worth of risk. Correct poker requires him to lay down his hand, rather than to call or raise. If B's estimate of the chance that A is bluffing is less than $\frac{5}{12}$, this is the correct procedure. If it is greater than $\frac{5}{12}$, he should either call or raise, preferably the latter, at least in case the player believes in his own judgments. Suppose that his estimate is that the chances are even that A is bluffing. B's expectation is now $\frac{1}{2} \times \$12$, or $6, so that he may look forward to a profit of $1. The reason that he may expect a profit with an even chance of winning is that there was $2 in the pot before his duel with A began.

This illustrates a very important point. It does not follow that because your chance to win is less than even, you should drop out of a hand. The correct statement is this: When your expectation is smaller than the amount that you are required to risk, you should drop out, and Hoyle's dictum concerning raising or retiring from the hand should be modified accordingly.

In these matters the player's expectation is all-important; his probability of winning the hand is important only through its influence on the expectation. The player who makes numerical judgments of the chances of each situation, and who bases his play on that principle, is certain to win against a player of equal judgment who does not. We have just seen in the above illustration that if B's probability is between $\frac{5}{12}$ and $\frac{1}{2}$, which means that the odds are against him on the hand, he will make a profit in the long run by calling. (It goes without saying that when the odds are with B he will make a still larger profit.) When the amount in the pot is large compared to the bet that the player is required to meet, this effect becomes very marked; in practice it is frequently advantageous for a player to call when his chance of winning the hand is only 1 in 4 or 5. But in each case he must be sure that according to the chances of the game and his best judgment of those chances that cannot be computed, his expectation is greater than the amount he must risk.

If other principles seem to work in practice, it is only because they are so widely used that almost everyone is in the same boat. In poker one does not play against par; one plays against other players.

CHAPTER *XII* ·

Roulette

ROULETTE is played, as the word implies and as everyone knows, with a wheel which can be spun rapidly and a small ball which finally comes to rest in one of the thirty-seven numbered compartments or slots at the edge of the rotating wheel. This ball or marble is caused to travel rapidly around a track, which slopes toward the center of the concave board where the wheel is rotating. On the inner edge of the track are set metal studs, so that the marble, after it has lost a part of its momentum, is certain to be deflected by one or more of them. The marble is impelled in the opposite direction to that in which the wheel is spinning, and the wheel is always set in motion before the marble.

These precautions are taken against the inevitable small bias due not only to mechanical imperfections, but to the difficulty of keeping the apparatus approximately level. A very small bias can produce a considerable effect in the long run. There can be no doubt, however, that at Monte Carlo every effort is made to keep the wheels as true as possible, so that each player will know what his chances are when he sits down to play, and it goes without saying that the casino tolerates no dishonesty on the part of employees. Indeed, the profits of the casino, except in bad times, have been so satisfactory that any deviation from this policy would be contrary to its every interest. We shall assume then, for the time being, that the roulette wheel is effectively without bias, so that the chance that a given number will turn up on a given *coup* is 1 in 37.

The chances in roulette vary according to the type of bet selected by the player. As played at Monte Carlo, the only form we shall consider, there are thirty-seven numbers, including the 0, and bets on any one of these (*en plein*), win thirty-five times the amount placed, so that the player receives from the croupier thirty-six times his stake. Bets can also be placed on

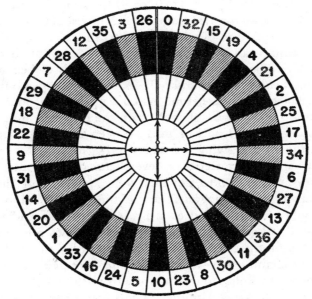

THE MONTE CARLO ROULETTE WHEEL.

two adjacent numbers by placing the stake on the line between them (*à cheval*), and if either number wins, the player's gain is seventeen times the amount staked. The next bet is on three numbers which form a row on the board (*transversale*), and this yields, when successful, a profit of eleven times the stake. The other bets, with the number of times the stake that can be won, are as follows: On four numbers (*un carré*), eight times; on six numbers (*sixain*), five times; on twelve numbers (*colonne*), two times; or on the first, second, or third dozen (*dou-*

zaine), also two times. There are in addition the three forms of bets on the "even" chances, which will be listed below.

Suppose the player places 100 francs on one of the numbers *en plein,* say on the 29. Let us compute his expectation, which is the amount he takes in if he wins multiplied by his chance of winning. The latter is $\frac{1}{37}$, the former 3,600 francs; his expectation is therefore $\frac{1}{37} \times 3,600$, or $97\frac{3}{10}$ francs, so that he is paying 100 francs for 97.3 francs' worth of risk. He loses in the long run 2.7 per cent of what he stakes; this is the gross profit that the casino can count on, applied to the total of all bets of this sort. Similarly, on each of the other kinds of bets on the numbers, *à cheval* for instance, the expectation comes out to be $\frac{1}{37} \times 3,600$ francs, as before. As far as profit and loss in the long run are concerned, then, bets on the numbers are equivalent to a game of heads or tails between the player and the casino, the former giving the odds of $51\frac{1}{3}$ to $48\frac{2}{3}$.

There are three varieties of "even" chances at Monte Carlo, and they are independent of each other, in the sense that the outcome of one does not affect that of the others. In all of them the number 0 is excluded as a direct win or loss, and the remaining thirty-six numbers are divided into two sets of eighteen, either on the basis of color *(rouge et noir)*, or on whether the winning number is even or odd *(pair et impair)*, or, finally, on whether it belongs to the numbers from 19 to 36, or to those from 1 to 18 *(passe et manque)*. The amount won or lost is the amount of the stake. Up to this point the game is fair; the odds are even. When the 0 comes up, a bet on the "even" chances is neither won nor lost, but goes to "prison," as one part of the table is called, where it stays until a number other than 0 appears. The player then wins back or loses what is in prison according to whether this *coup* is favorable or the opposite. In other words, when the 0 comes up he has very nearly 1 chance in 2 of getting back his money, and very nearly 1 chance in 2 of losing it; he has no chance of winning on the play.

We shall compute the expectation of a player who places 100 francs on one of the even chances at Monte Carlo. There are thirty-seven possible results of which eighteen are favorable, so that his probability of winning is $\frac{18}{37}$. If he does not win he has still a chance of getting back his stake, which will happen if the 0 comes one or more times, followed by a favorable *coup;* the probability that this will happen is $\frac{1}{74}$. His expectation is therefore $(\frac{18}{37} \times 200) + (\frac{1}{74} \times 100)$, or 98.65 francs. On the even chances, then, the player loses 1.35 per cent of what he stakes, in the long run, as compared to 2.7 per cent on the numbers, in other words one half as much. This fact induces many "serious" players, who are content to aim at relatively modest winnings, to stick to the even chances. They correspond to a game of heads or tails in which the player gives the bank odds of 50⅔ to 49⅓.

A remarkable feature of roulette at Monte Carlo is that the 0 is not an outright loss on the even chances. We have just discussed the effect of the "prison" system and have seen that it reduces the odds against the player in such a way that he loses only 1.35 per cent of what he stakes, in the long run, as compared with a loss of 2.7 per cent, if he plays on the numbers, or on any combination of numbers. In some American gambling houses, where roulette is played, there are actually two zeros, and if either of them comes up, the player immediately loses his stake. This means that there are 2 chances in 38, or 1 in 19 that the player loses his bet. In the long run he loses 5.26 per cent of the total stake, as compared to 1.35 per cent in Monte Carlo roulette, on the even chances.

The casino has seen fit to offer this inducement to play on the even chances. I do not know the reason for this difference, but my guess would be that it is because almost all systems are played on the even chances, and the casino welcomes system players with the same enthusiasm that most of the latter welcome the opportunity to play. The result is that there probably exists nowhere else a public gambling game with such small odds in favor of the house.

We have been assuming that the roulette wheel is free of bias and that the croupier operates it in a perfectly honest manner. As far as the latter condition is concerned, one can only say that it is impossible to prevent occasional dishonesty on the part of individual croupiers. The consensus of those who play much is, however, that at Monte Carlo such an occurrence is very rare.

On the matter of the extent to which it is possible for the croupier to influence the result of a *coup,* opinions differ widely. It is not at all necessary, in order to produce wins for accomplices, that he should be able to cause a certain number to come up, but merely that he exert enough influence on the play to cause several adjacent numbers to come up a little more frequently than the others, in the long run. Furthermore, it is sufficient for the croupier to have knowledge of the bias of the apparatus at the time in question. In the matter of this bias it is very probable that every roulette wheel is slightly affected by it. But it is also very probable that this bias changes from day to day; if it stayed the same for long series of *coups,* it could easily be detected by statistical methods, and observant players might well be able to place their bets in such a way that the odds were actually in their favor. The effect on the casino's profits, if there were many observant players, would be enormous.

What we have just said applies to bets on the numbers, where it is possible to put money on several chances that are actually adjacent on the wheel. This does not by any means apply to the even chances, as everyone who has ever examined a roulette wheel knows. The red and black alternate, half even and half odd; half are over 18, the other half under 19. It is rather difficult to imagine a technique for dishonest manipulation, which does not mean that it cannot be done.

Any sustained dishonest manipulation, or any bias large enough to have a pronounced effect on the odds, could surely be detected by statistical means.

In 1894 Professor Pearson published a statistical study of the even chances in Monte Carlo roulette (*Fortnightly Review,* February, 1894), based on about twenty-four thousand *coups,* which led him to the following conclusions: Playing odd and even is equivalent to playing heads or tails (the 0 is left out of account). In red and black the two chances come up the same number of times in the long run, as in heads or tails, but the runs (see Chapter VI) do not appear to follow the same law as they do in the latter game. "Short runs are deficient, and the color changes much more frequently than the laws of chance prescribe."

It is worth noticing that these observed variations from the laws of chance occur where the two chances alternate on the wheel, while in odd and even, where they are more irregular, no such variation is observed. I am not aware of any later confirmation of these results and have not sufficient data on the manufacture and testing of the apparatus to be able to say whether or not these conclusions are still applicable after the lapse of more than fifty years. In any case we shall assume that the even chances operate, apart from the effect of the 0, after the pattern of heads or tails. That this is true of *pair et impair* and of *passe et manque* has, I believe, never been seriously questioned, and if we use the language of *rouge et noir,* anyone so inclined can easily translate it into that of the others.

The play on the numbers contains very little of particular interest to us. Owing to the large odds against winning on any particular *coup,* and to the relatively low limit on the amount of bets, systematic play is practically out of the question. It would be possible to play the systems on *colonne,* or on *douzaine,* where the chance of winning is 1 in 3, but as the game is less favorable to the player than those on the even chances, this is not often done. Bets on the numbers give the fullest play to "runs of luck" and the reverse, as statistical order does not become apparent unless the number of *coups* considered is very large, several thousand, at the very least. Tales of curiosities and freaks concerning the behavior of the numbers are legion and are usually much less astonishing than the absence of such sports would be over a long period. A number has appeared three or four times running, or has not appeared for several hundred *coups*, or fifteen to twenty alternations between red and black or odd and even have occurred.

The chance that any given number will come up on the next *coup* is 1 in 37, and the chance that a number *written down in advance* will come up twice running is therefore 1 in 1,369. If the number is *not* designated in advance, and we ask the chance that the next two *coups* result in the same number, the chance is 1 in 37, and the chance that there will be a run of 3 on the same number is 1 in 1,369. The chance that there will be a run of 4 is similarly 1 in 50,653, and as each table averages over three thousand *coups* per week, such an event is to be expected a few times each season.

In a series of one hundred *coups* the chance that a given number will not come up at all is about 1 in 15; the odds are 14 to 1 that the number will come up one or more times in the series. The chance that it will come up exactly once is 1 in $5\frac{1}{2}$, so that the odds in favor of its appearing at least twice are 3 to 1. In a series of ten *coups* the odds are 3 to 1 that the number will not appear one or more times, and 32 to 1 against two or more appearances. If the bet is on two numbers (*à cheval*), the results are quite different. The bet is won if either or both appear, and there is no need to distinguish between the two cases. The chance of winning is $\frac{2}{37}$ on each *coup*, and the odds against winning one or more times on the series of ten *coups* are almost exactly 4 to 3. The odds against winning twice or more are about 9 to 1.

Play on the even chances, when the effect of the 0 is neglected, and the wheel is assumed to be mechanically perfect and the play fair, is equivalent to the game of heads or tails, which we have already discussed in Chapter VI. There we gave an example of one of the many systems that are in use, the Martingale, and saw that, due to the limits on bets placed by the bank, the player is certain to encounter adverse runs that must be interrupted, leaving large losses to be recouped one unit at a time. Still confining ourselves to heads or tails, we can sum the matter up as follows: The expectation of the player is exactly equal to the amount he stakes (that is, the game is fair) and no amount of juggling with the order and amount of bets, or with the choice of chances, can change this fundamental fact. Therefore the player does not in the least change his prospects *in the long run* by playing a Martingale or any other system.

There is one other factor that must not be overlooked. We saw in Chapter VIII that when the chances of the game are even, the odds in favor of the player's losing his entire capital, if the play is continued, are in the proportion of the bank's capital to his own. If his own available capital is only a small fraction of the bank's, his chance of losing it is very large. If he is playing a Martingale, he will sooner or later encounter a series of maximum losses so close together as to leave him no opportunity to recoup his losses in between.

If we leave the game of heads or tails and turn to the even chances of roulette, the player's position becomes considerably worse. For the odds are now slightly against him, and we have seen in Chapter VIII that under these conditions even a large amount of capital cannot save him from ultimate ruin. It remains true, nevertheless, just as it was for heads or tails, that *for a short enough period of play,* his chance of a small win with the Martingale is large, his chance of a large loss small. These risks exactly compensate each other, in the sense that the player's expectation is always 98.65 per cent of what he stakes. And by substituting other systems for the Martingale he can shift these chances about almost as he pleases; for instance by playing an "anti-Martingale" (see Chapter VI) he will have a small chance of a large profit, and a large chance of a small loss. What he cannot change is the value of his expectation, which remains the same for any system of play.

The effect of the 0 in the Martingale (or in any system where the amount placed progresses with a loss) is not merely to make the odds slightly against the player; though it is fatal in the long run, it is relatively unimportant in a very short series of *coups*. It also tends to increase the unfavorable runs, which are all too frequent to suit the player's taste at best. To offset this additional "strain," some players place "insurance" on the 0. It is done in some such manner as this: After four consecutive losses the player is 150 francs out of pocket, his initial bet of 10 francs having been doubled three times, while the Martingale requires him to place a bet of 160 francs on the succeeding *coup*. At this point he places 10 francs *en plein* on the 0. If the 0 comes, he wins 350 francs at once, and his 160 franc bet on the even chances goes to prison. If the next *coup* is also unfavorable, he is nevertheless 40 francs ahead on the series; if it is favorable, he gets his 160 francs back, and is 200 francs ahead on the series. If the 0 does not come on the fifth *coup* of the series, and if this last is a loss, requiring him to bet 320 francs on his color, he backs the 0 again, this time placing 15 or 20 francs on it, and so on, always placing enough so that if the 0 does come, he can break off the series with a profit, or at least with a much reduced loss, and start again.

The effect of these bets on 0 is that on the average the player loses 2.7 per cent of the total so staked, thus increasing very slightly the odds against him of the game as a whole, while in return for this "insurance premium," he gains the advantage of a lessened strain on the unfavorable runs. By properly scaling these bets on the 0 it is possible for the game, in so far as the unfavorable runs are concerned, to be on an even basis with heads or tails.

All these systems of the Martingale species are essentially

alike; they consist of a progression of stakes, more or less rapid, and a set of rules to tell you when to progress. By varying these last you can bet on or against long runs, or short runs, or inter-mittences (changes of color), or you can arrange them so that you will win a large amount every time there is a run of four blacks followed by three reds, or some other arbitrary pattern. By varying the progression you can change the tempo of the game; the more rapid the progression, the greater the risk of being forced to the maximum permissible bet, but also the greater the chance of a small profit in a short game. The pos-sible variations on these themes are almost unlimited; hun-dreds have been published and one hears of others too valuable for the public eye. In fact, the writer has had several letters from individuals concerning the systems they have invented which they would disclose only under promise of complete secrecy. Since all possible systems are combinations of a few basic ideas, they are, in a sense, known, and such a promise could not well be given.

If you are going to Monte Carlo, or to any casino where the play is similar, and must play roulette or *trente et quarante* with a system, by all means make your own. It is necessary to understand only the few basic principles at the root of all of them and you can turn them out by the score. In that way you can have a system that suits your personality. I refer to systems of the sort discussed above, which have a definite effect on the player's chances in a short series of *coups,* not to those based on astrology, numerology, or on some other form of superstition. The latter require considerably more ingenuity of workman-ship, but they have no effect on the game, one way or the other.

There is one and only one system of play that almost surely leads to winnings in the short run, but the amounts won may be small. Curiously enough, this system depends on finding a dishonest casino. If you have discovered a roulette wheel that is being manipulated, you can be quite sure that it will be manip-ulated in such a way that whatever large bets are placed will likely be lost. If there is a large amount bet on the even numbers, for example, you place a much smaller amount on the odd numbers. If you are right about the dishonest manipulation, the odds on such bets are in your favor, and if you continue long enough you are certain to win. This system, if that is the right name for it, has been successfully executed in at least one case.

One of the systems commonly played and known as the Labouchère, which involves a less dangerous progression than the Martingale, runs as follows: One bets consistently on red or black or even, or changes to suit his fancy (it makes absolutely no difference), having first written on a sheet of paper a few numbers in a column, the first ten numbers, for example, although any set whatever will do. The first bet is the sum of the top and bottom numbers, 10 + 1 in this case. If it is won, these numbers are scratched out, and the next bet is the sum of the top and bottom numbers of the remaining column, 9 + 2 in this case. If it is lost, the amount of the loss is written at the end of the column. The next bet is always the sum of the first and last numbers. When and *if* all the numbers are scratched off, the win amounts to the sum of all the numbers in the original column, 5×11, or 55, in the example given.

The Martingale, as described in Chapter VI, wins one unit for every favorable *coup*, or about ½ unit for every *coup*, as long as no unfavorable run of too great length is encountered. (It is assumed that play is discontinued after a favorable *coup*.) The progression used after a loss is 1, 2, 4, 8, 16, etc. If the progression 1, 3, 7, 15, 31, 63, etc. (always one less than a power of 2), is substituted, the game is known as the Great Martingale. With this system, when no disastrous runs are met with, the player wins one unit (10 francs) for every spin of the wheel, less losses on the 0, if any.

The results are equivalent to those obtained if two partners play on opposite sides of the table, one betting consistently on red, the other on black, each playing an ordinary Martingale.

The point of this partnership game is that if there is a long run on one of the colors, black for instance, the red player is undergoing a considerable "strain," the black player none, so that the latter can temporarily (at least) transfer his capital to the other. Each in turn can play the role of the good Samaritan. That is perhaps the greatest virtue of this partnership system; for, as just stated, one person could conduct the game quite as well as two if he had the use of his partner's capital. When you see two partners apparently playing against each other, you must keep in mind that their bets are always unequal. If one of them placed a bet equal to the difference between the two bets (on either color), the net result would be the same, in the end, as that of the partnership game, and this leads to the identical progression used in the Great Martingale. I was once watching a game of this sort in progress when a Frenchwoman nearby turned to her escort and said, "Look at those imbeciles over there. They are evidently partners, and one plays the black, the other the red. Whenever one wins the other loses."

In discussing the Martingale system we have remarked several times that in playing it the chance of a small win on a *short* series of *coups* is large, while there is a small chance of a large loss. The use of such a word as "short" is vague, and the reader has the right to ask what is meant by it. Does it mean one hundred *coups*, a thousand, ten thousand? We shall answer by giving the actual figures for a series of one thousand *coups*. The player bets always on red, his initial stake being 10 francs, which he doubles after each successive loss.

If he loses ten consecutive *coups*, his next stake is 10,240 francs, and as the limit is 12,000 francs, he is not permitted to double again if he loses the *coup*, and must accept his loss of 20,470 francs. We shall say, then, that a "disastrous run" is one of eleven black. The probability of such an occurrence is (Chapter VI) $1/2^{13}$, or 1 chance in 8,192, and the probability of a run of eleven *or more* black is $1/2^{12}$, or 1 chance in 4,096.

The chance that this disaster will overtake the player one or more times in a series of one thousand *coups* comes out to be 22/100. The odds that it will not happen are 3½ to 1.

If we considered a game of five hundred *coups*, which is about a day's play at a table, instead of one of a thousand, the odds in favor of the player (that is, the odds against a disastrous run) would be 7.3 to 1. The 0 is left out of account, but its effect on the odds would be very slight. To sum up: In a short game of five hundred *coups* it is 7 to 1 that no adverse run of 11 or more will be encountered, and therefore that the player will win about 250 units or 2,500 francs (it is assumed that the player's unit bet is the minimum of 10 francs; otherwise his chance of encountering an adverse run is greater). If he were to continue to ten thousand *coups,* the odds would be reversed and would become 11 to 1 that he *would* run into a disastrous series. If he were fortunate enough to escape such a series, his winnings would amount to something like 50,000 francs.

When a system of the type of the anti-Martingale is played, these conclusions are reversed. The player leaves his stake, or a part of it, when he wins a *coup,* and does not withdraw stakes or winnings until he has won a certain number of times in succession, or he may even add to them after each win. His policy in this respect, whether he leaves stakes and winnings to accumulate (this is called a *paroli*), or whether he withdraws a part or adds to them, determines the "progression," which is correspondingly rapid or slow. The larger the number of successive wins before profits are taken out of play, the greater the amount that it is possible to win.

Suppose that an anti-Martingale is played, and that stakes are allowed to accumulate until there have been eleven consecutive wins (*paroli* of 11). This is the longest *paroli* permitted by the Monte Carlo limit. The outcome for the player is as follows: On a favorable run of 11 or more he wins 2,047 units, or 20,470 francs; on a favorable run of less than 11 he breaks even, while he loses one unit each time that the unfavorable color appears. His average loss in the long run (if there is no favorable series), is therefore ½ unit per spin of the wheel. The situation is the exact reverse of that in the ordinary Martingale, and the player's chances are the reverse of those determined above. In a series of one thousand *coups* the odds are 3½ to 1 that there will be no run of 11 or more on the favorable color or, in other words, that the player will lose 250 units. If he does encounter one such run, the amount that he wins is not far from 2,047 minus 250 units, or 1,797 units, roughly 18,000 francs. If the game continues, there will certainly be a run of 11 or more; on the other hand, the losses of one unit accumulate at just such a rate that in the end the player loses 1.35 per cent of the total amount staked. If "insurance" is placed on the 0, this loss is increased by 2.7 per cent of the amount so placed.

Before leaving the subject of roulette there is one other point that deserves a remark in passing. If the apparatus is mechanically perfect, it is a matter of complete indifference whether a bet is placed on one number or another, on red or on black, on odd or on even. But if the apparatus is defective, this ceases to be true. Suppose, for example, that "short runs are deficient, and the color changes much more frequently than the laws of chance prescribe," to quote Professor Pearson once more. Then it is easy to devise a system that will take advantage of this fact, and if the variations from the laws of chance are sufficient, it may turn out that the odds of the game desert the bank and come over to the player. It would be necessary only to place the bet in each instance on the opposite color to the one that showed on the preceding *coup*. Any progression could be used, and the game could be conducted either in the Martingale or the anti-Martingale tradition.

Clearly it is to the interest of the casino that the laws of roulette coincide with the laws of chance.

CHAPTER XIII

Lotteries, Craps, Bridge

LOTTERIES are a simple illustration of chance. A lottery with cash prizes, as distinguished from a raffle, is conducted essentially along the following lines: Suppose that there are one million chances, each selling for $1, and that all the chances are sold. Suppose that there are three prizes, the first $300,000, the second $200,000, and the third $100,000. Each person who purchases one ticket has three chances in 1,000,000 to win one of the prizes. His expectation, as defined in Chapter VII, is the measure of his chance of success. Since one ticket can win but one prize, his expectation is

$$\frac{300,000}{1,000,000} + \frac{200,000}{1,000,000} + \frac{100,000}{1,000,000} \text{ or } \frac{600,000}{1,000,000}.$$

Looked at as a bet, the ticket holder has odds against him of 6 to 4. The difference is represented by the $400,000 made by the promoters. If he purchases two tickets, his chance to win first prize becomes 1 in 500,000, that to win one of the prizes, 3 in 500,000; the odds against him remain the same. In spite of the unfavorable odds, buying lottery tickets is attractive to many people, because the price of the ticket is so small, the prizes so large.

Lotteries of this sort are illegal in the United States, in the sense that they are forbidden the use of the mails. There are also laws against them in some states. However, lotteries are still conducted in this country without the use of the mails, to raise money for public and charitable institutions.

There is a wide difference of opinion as to whether such lotteries should be permitted. Many people object to them, regardless of where the proceeds go, on the ground that they are gambling. Some leading sociologists object to them on the ground that the sudden acquisition of a large sum of money by an individual who has not worked for it has bad social consequences. On the other hand, many people approve of such lotteries when the proceeds are used for commendable purposes.

Few people know that lotteries played an important part in the early history of the United States. Funds to carry on the Revolutionary War, to the extent of $5,000,000, which was a lot of money in those days, were obtained from lotteries. It is said that George Washington purchased the first ticket. In 1655 the city of New Amsterdam, now known as New York, raised money for the benefit of the poor by running a lottery. Prizes to the winners were Bibles. On several occasions Harvard College obtained funds by similar means. Lotteries in this country came into ill repute in the latter part of the nineteenth century due to flagrant abuses.

In most civilized countries, except the United States and England, lotteries are permitted, at least when conducted by the government or by a municipality. Here are some facts about lotteries in other countries that may be of interest: They date back to the days before World War II, when conditions were more normal. In France there were lotteries conducted by the city of Paris and by the Crédit Foncier, a semiofficial mortgage bank. There was also a lottery with five main prizes of 5,000,000 francs each (about $133,000 at the rate of exchange then prevailing) and one hundred other prizes of 1,000,000 francs each. The prizes represented 60 per cent of the proceeds, the other 40 per cent going into the treasury. Of the 40 per cent, 100,000,000 francs was set aside for agricultural relief.

In Spain the government lottery was held at Christmas time. The 10,000 prizes totaled 15,000,000 pesetas, the profits going to the government. In Italy there were both state and municipal lotteries, many of them held as often as once a week. In 1933 Mussolini held a lottery to raise money for the electrification of the Italian railroads. New South Wales, Australia, in 1931 legalized a state lottery. The proceeds were turned over to hospitals. In 1933 the Nanking government of China established a $5,600,000 lottery. Of this amount, $2,800,000 was used to build new roads and purchase airplanes. The surplus from the Swedish state lottery goes to Red Cross hospitals, museums, and to the support of music, art, and drama.

One of the most popular out-and-out betting games seems to be the game of craps. The rules of this game are extremely simple. It is played with two ordinary dice. The man who has possession of them wins immediately if the total of the spots on his first throw is 7 or 11. He loses immediately if this total is 2, 3, or 12 but continues to throw the dice. If the total is any one of the remaining six possible points, he neither wins nor loses on the first throw, but continues to roll the dice until he has either duplicated his own first throw, or has thrown a total of 7. The total shown by the dice on his first throw is called the crapshooter's *point*. If he throws his point first, he wins. If he throws 7 first, he loses and is required by the rules to give up the dice.

There are several interesting questions to be asked in connection with this game, from the standpoint of chance. We ask first whether the player throwing the dice has the odds with him or against him. As it is well known that when craps is played in gambling houses, the house never throws the dice, and as we know that the odds are always with the gambling house, we may feel reasonably certain in advance that our calculations, if correct, will show that the odds are against the player with the dice.

Other questions can be asked concerning the odds in favor of or against making certain throws. For example, we can ask the odds against throwing a total of 6 before throwing a total of 7. This particular problem is part of the broader one of determining the odds against the player who throws the dice.

In Chapter V, among our illustrations of basic probability principles, we considered the various chances in throwing two dice. We obtained the following table of probabilities for the various throws:

(TABLE III)

Total of Throw	Probability
2 or 12	$\frac{1}{36}$
3 or 11	$\frac{2}{36}$ (or $\frac{1}{18}$)
4 or 10	$\frac{3}{36}$ (or $\frac{1}{12}$)
5 or 9	$\frac{4}{36}$ (or $\frac{1}{9}$)
6 or 8	$\frac{5}{36}$
7	$\frac{6}{36}$ (or $\frac{1}{6}$)

With this table and the rules of the game before us, we see at once that the probability that the player throwing the dice will win on his first throw is $\frac{8}{36}$, or $\frac{2}{9}$. The probability that he will lose is $\frac{4}{36}$, or $\frac{1}{9}$. The chance that he will win or lose on the first throw is therefore $\frac{3}{9}$, or $\frac{1}{3}$. The probability that the first throw will not be decisive is $\frac{2}{3}$.

We have now disposed of the cases where the first throw yields one of the following points: 2, 3, 7, 11, 12. Suppose now that some other total results from the first throw, for definiteness let us say 6. Six, then, becomes the player's point. In order to win, he must throw a 6 before he throws a 7. What is his chance of doing so? The chance of throwing 7 is $\frac{6}{36}$, that of throwing 6, $\frac{5}{36}$. One might be tempted to reason as follows: The ratio of the chances of throwing a 7 to those of throwing a 6 is 6 to 5. Therefore the probability that a 6 will appear before a 7 is $\frac{5}{11}$. The probability that 7 will appear first is $\frac{6}{11}$.

This is an example of bad reasoning which gives the correct result. It is bad reasoning because, as we have stated it, the conclusion does not follow from the principles of probability.

Also, we have assumed that either a 6 or a 7 is certain to appear if we roll the dice long enough.

Let us attempt to make this reasoning correct. The player's point is 6. The probability that he will throw a neutral point (all throws except 6 and 7) is $25/36$. The chance that he will make two consecutive neutral throws is $(25/36)^2$, and the probability that he will make this throw n times in succession is $(25/36)^n$. For the game to continue indefinitely, it would be necessary for the player to throw an indefinitely large number of neutral throws in succession. But the probability of doing so is $(25/36)^n$, which becomes smaller and smaller as n increases. Since we can make it as small as we please by taking n large enough, we can legitimately consider the probability that the game will not end as 0.

With all neutral throws thus eliminated, there remain to be considered only the throws of 6 and 7. We can now conclude that out of the eleven possible cases that give a 6 or 7, five favor the 6 and six favor the 7. Therefore the probability of throwing a 6 before throwing a 7 is $5/11$.

It is easy to make the corresponding calculation for each of the six possible points. The probability is the same for the point 6 as it is for 8, the same for 5 as for 9, and so on, just as in the preceding table. The calculation for each of the possible points gives the following results:

TABLE XII

PROBABILITY OF THROWING INDICATED POINT
BEFORE THROWING A SEVEN

Points	Probability
4 (or 10)	$3/9$
5 (or 9)	$4/10$
6 (or 8)	$5/11$

We wish to know the probability, before the first throw in the game is made, that the crapshooter will win on each of the points just listed. In order to win in this manner he must, of course, neither win nor lose on his first throw. We find what we wish by combining the two preceding tables as follows:

TABLE XIII

PROBABILITY OF WINNING ON INDICATED POINT

Points	Probability
4 (or 10)	$\frac{3}{36} \times \frac{3}{9} = \frac{1}{36}$
5 (or 9)	$\frac{4}{36} \times \frac{4}{10} = \frac{2}{45}$
6 (or 8)	$\frac{5}{36} \times \frac{5}{11} = \frac{25}{396}$

This means that the probability that the crapshooter will win on a point specified in advance, say point 5, is $\frac{2}{45}$, and the probability that he will win on point 9 is also $\frac{2}{45}$.

To find the *total* probability that the crapshooter will win, we add to his probability of winning on the first throw, which is $\frac{2}{9}$, the sum of the three probabilities shown in this table, each multiplied by 2. This gives a probability of 244/495, or 0.49293.

This is the sort of result we expected to begin with. The odds are against the crapshooter, although by only a very small margin. In fact, in the long run, the crapshooter loses (or the gambling house wins, if there is one) only 1.41 per cent of the amounts staked. This compares with the loss of 1.35 per cent for the player of the even chances at Monte Carlo roulette (see Chapter XII), and 2.7 per cent on the numbers in roulette.

A great deal of the interest in the usual crap game comes from side bets of various sorts, made between the players. Many of these bets consist in giving odds that one total will appear before another. One player might say to another, for example, "Two to one that a 6 appears before a 4."

The problem of finding the fair odds in such cases was solved when we found the odds against the crapshooter. The fair odds on all such bets, including those totals that cannot be a player's point, by the rules of the game, can be put in the form of a table. This table gives the odds against throwing one or other of the totals given in the left-hand column before throwing one or other of the totals in the top line. The listing together of two numbers, such as 4 and 10, is done only to abbreviate. The odds as given apply to either 4 or 10, not to both on the same series of throws.

TABLE XIV

Odds against Throwing	Before Throwing				
	3 (or 11)	4 (or 10)	5 (or 9)	6 (or 8)	7
2 (or 12) . . .	2 to 1	3 to 1	4 to 1	5 to 1	6 to 1
3 (or 11)		3 to 2	4 to 2	5 to 2	6 to 2
4 (or 10)			4 to 3	5 to 3	6 to 3
5 (or 9)				5 to 4	6 to 4
6 (or 8)					6 to 5

Thus the correct odds against throwing a 4 before throwing a 6 are 5 to 3. Otherwise stated, the odds in favor of throwing the 6 first are 5 to 3. The player who said "Two to one that a 6 appears before a 4" is therefore making a bad bet. In the long run he will lose $\frac{1}{24}$ of the amount staked.

The game of contract bridge, although it involves a large element of skill, offers some interesting examples of the workings of chance. In fact, compared to the purely intellectual game of chess and the oriental Go, one of the outstanding features of bridge is precisely this chance element in the strength of hands and in the distribution of cards, which leaves full sweep to runs of luck, and which provides an important part of the attraction of the game.

One of the more interesting bridge problems that involves the theory of probability concerns the relation between bidding and the score. There are many games in which the score exercises a decisive influence on sound play, and bridge is certainly one of them. If, in the face of almost certain loss, you make a bid that actually loses 300 points but prevents your opponents from making a bid that would score 700 points for them, you have a net gain of 400 points. To the extent that your judgment of the alternatives was correct, this is certainly sound play and is so recognized by the leading authorities. Some people have difficulty in appreciating this fact, since if you lose 300 points, that is the only item that appears on the score sheet. You had, in fact, the choice of losing 300 points or of losing 700 points. There was only one thing to do.

Problems of this sort are best approached through the idea of mathematical expectation, which we studied in Chapter VII. We recall that the idea of expectation makes possible a sort of probability bookkeeping, according to which credit and debit entries are made on future events, much as though they had already taken place.

After one side has scored the first game in a rubber, it has gained a very decided advantage. Its probability of winning the rubber is greatly increased. It has, in a sense, earned a part of the rubber premium, and in the case of unfinished rubbers the rules of the game allow for this fact. We shall therefore look at this partial premium as an invisible score that does not appear on the score sheet. We wish to find the value of this intangible item. Let us assume that North-South has just made the first game, and that the two sides are of equal strength so that, in the long run, each will win an equal number of games. What is the probability that North-South will win a second game, and therefore a rubber, before East-West has won a game? The answer is clearly $\frac{1}{2}$, since the probabilities of the two sides to win the next game are equal and since we can assume that one side or the other will win the next game. (Compare with the similar situation in the game of craps, discussed earlier in this chapter.)

We now ask the probability that North-South will win the rubber *after* East-West has won the next game. The probability that East-West will win the next game is $\frac{1}{2}$, and the probability that North-South will win the next succeeding game is also $\frac{1}{2}$. The required probability, being the product of these two probabilities, is therefore $\frac{1}{4}$. Finally, then, the probability that North-South will win the rubber, on the assumption that North-South has won the first game, is $\frac{3}{4}$.

We wish now to find the expectation in points that North-South is justified in entering on the score sheet (in invisible ink), as a result of winning the first game. The rubber premium, if rubber is made in two games, is 700 points. The probability that North-South will accomplish this is $\frac{1}{2}$. Their point expectation is therefore $\frac{1}{2} \times 700$, or 350. If North-South makes the rubber in three games, they will score a premium of 500 points, and the probability of their doing this is $\frac{1}{4}$. This gives an additional expectation of $\frac{1}{4} \times 500$, or 125 points. The total point expectation is therefore 475. The point expectation, as so computed, omits honor points and possible overtricks.

This is the credit side of the ledger. How about the debit side? The probability that North-South will lose the rubber is $\frac{1}{4}$, and if so the rubber premium is 500 points. Their expectation of loss is therefore $\frac{1}{4} \times 500$, or 125 points. Their net expectation, which is the same as the "invisible item" on the score, is the difference between their expectations of winning and of losing. This difference is 475 minus 125 points, or 350 points. It is a credit to the account of North-South.

We can look at this calculation in a slightly different but equivalent way. North-South has an expectation of winning 475 points. East-West has 1 chance in 4 of winning the rubber, with a premium of 500 points, giving an expectation of winning of 125 points. (This must agree, as it does, with North-South's expectation of loss.) Now if we considered that North-South has an invisible item of 475 points, and that East-West has an invisible item of 125 points, it would evidently come to the same thing to say that North-South has a net invisible item of 350 points. The rules of bridge call for a premium of 300 points for the winner of the first game of an unfinished rubber. This apparently should agree with the net expectation of the side that has won the first game. But it does not agree, for we have just found that its value is 350 points.

The total expectation of winning of North-South and East-West combined is 475 points plus 125 points, which equals 600 points. This is clearly the correct value of the total expectation, since there is 1 chance in 2 that 700 points will be won, and 1 chance in 2 that 500 points will be won. The sum is the average of 700 and 500, or 600.

We have assumed that North-South and East-West are teams of equal strength. It is interesting to see what happens to the point expectations if we make a different assumption. Let us assume that the probability that North-South will win is p_1, and the probability that East-West will win is p_2, so that $p_1 + p_2 = 1$. Then if North-South has already won the first game, the probability that North-South will win the rubber is

$$1 - \left(\frac{p_2}{p_1 + p_2} \right)^2 = 1 - p_2^2.$$

For example, if $p_1 = \frac{3}{5}$ and $p_2 = \frac{2}{5}$, the probability that North-South will win the rubber is found by substituting the value of p_2 in the above expression, giving as result $2\frac{1}{25}$. Thus, as would be expected, the extra strength of North-South leads to an increase in the probability of winning the rubber, amounting to the difference between $\frac{3}{4}$ and $2\frac{1}{25}$. It should be noted that if p_1 and p_2 are each equal to $\frac{1}{2}$, the value of the above expression is $\frac{3}{4}$, as it should be.

The probability that North-South will win the rubber in straight games is p_1, giving a point expectation of 700 p_1 points. Their probability of winning the rubber with East-

West also vulnerable is p_1p_2, and the corresponding expectation is 500 p_1p_2. The probability that North-South will lose the rubber (or that East-West will win it) is p_2^2 and the corresponding expectation is — 500 p_2^2 (note the minus sign). The total expectation, call it E, which equals the invisible item on the North-South side of the score, is therefore

$$E = 700 \, p_1 + 500 \, p_1p_2 - 500 \, p_2^2.$$

Remembering that $p_1 + p_2 = 1$, this can also be written

$$E = 2{,}200 \, p_1 - 1{,}000 \, p_1^2 - 500.$$

If $p_1 = \frac{3}{5}$, as in the illustration used before, we find that E = 460. The greater playing strength of North-South has increased the invisible item on their score by 110 points. We notice also, as a check on our reasoning, that if $p_1 = \frac{1}{2}$, E = 350, as it must.

We could equally well have written the formula for E entirely in terms of p_2, instead of p_1. In that case we find that

$$E = 700 - 200 \, p_2 - 1{,}000 \, p_2^2.$$

We get the same value for E whichever of these expressions we use, provided only that the values of p_1 and p_2 add up to unity.

A hand in bridge that contains no face card or ace—in other words, a ten-spot-high hand—is called, I believe, a Yarborough. Apparently there is some confusion as to which hand should properly be called by this name. In his *Chance and Luck*, Richard A. Proctor gives its origin as follows: A certain Lord Yarborough once made it a habit to offer bets of 1,000 to 1 against the occurrence of a whist hand containing no card above a nine spot. Apparently this would indicate that a Yarborough means a nine-spot-high hand, while we have called a Yarborough a ten-spot-high hand, following current practice.

On the subject of such wagers, Proctor says: "Odds of a thousand pounds to one are very tempting to the inexperienced. 'I risk my pound,' such a one will say, 'but no more. And I might win a thousand.' That is the *chance;* and what is the *certainty?* The certainty is that in the long run such bets will involve a loss of 1,828 pounds for each 1,000 pounds gained, or a net loss of 828 pounds. As certain to all intents as that 2 and 2 make 4, a large number of wagers made on this plan would mean for the clever layer of the odds a very large gain." *

The chance of holding a nine-spot-high hand is correctly stated above as 1 in 1,828. The chance of a ten-spot-high hand is 1 in 274.8, a result which is easily computed as follows: Imagine, for simplicity, that your hand (the one in question) is dealt from the top of the deck. There are thirty-six cards below the rank of jack, from which the thirteen cards in your hand must be selected, in order for the hand to contain no face cards or aces. The chance that the first card will be one of these thirty-six is therefore $36/52$. Assuming that the first card is "favorable," the chance that the second card will also be favorable is $35/51$. The chance that both the first two cards will be favorable is therefore $\dfrac{36 \times 35}{52 \times 51}$. The chance that all thirteen cards will be favorable, thus giving a ten-spot-high hand, is $\dfrac{36 \times 35 \times \ldots \times 24}{52 \times 51 \times \ldots \times 40}$. This is equal to 1/274.8, as previously stated. If you multiply this result by $\dfrac{23 \times 22 \times 21 \times 20}{36 \times 35 \times 34 \times 33}$, you will have the chance that your hand will contain no card above the nine spot. These are the hands on which Lord Yar-

* *Chance and Luck,* Richard A. Proctor, Longmans, Green and Co., Inc., New York, 1887, p. 79.

borough is reputed to have made his wagers, and the chance of holding such a hand is 1/1,828. If this last result is multiplied by $\frac{19 \times 18 \times 17 \times 16}{32 \times 31 \times 30 \times 29}$, you will have the chance of holding an eight-spot-high hand. This comes out to be 1 in 16,960. Similarly, the chance of a seven-spot high is 1 in 254,-390, that of a six-spot high 1 in 8, 191, 390. You have probably never held one of the latter. You seldom see one of these latter hands in actual play.

There are other simple chances in bridge that are of some interest. One is the following: If the dealer and dummy hold the ace and king of a suit of which the opposing hands hold four, including the queen, what is the probability that the queen will fall if both the ace and king are led? In other words, what is the probability that these four cards will be divided equally between the two opposing hands, or that the queen will be alone in one of them? The calculations are easily made and give the following results: Call the two opposing hands North and South. Then the possible distributions of the four cards, together with their respective probabilities of occurrence are given in the following table:

TABLE XV

North	South	Probability
4	0	110/2,300
0	4	110/2,300
3	1	572/2,300
1	3	572/2,300
2	2	936/2,300

The most probable distribution is therefore an equal division. If, however, we consider only the division of cards between the two opposing hands, regardless of how many are in North or in South, this result is changed. We then obtain the following table:

TABLE XVI

Distribution	Probability
4 and 0	220/2,300
3 and 1	1,144/2,300
2 and 2	936/2,300

Now the most probable distribution is seen to be three cards in one of the opposing hands, and one card in the other. The queen will fall on the lead of ace and king if the distribution is two and two, or if the distribution is three and one, the single card representing the queen. Assuming the distribution of three and one, the probability that the one card is the queen is $\frac{1}{4}$. The probability that the queen will be alone in one of the hands is therefore $\frac{1}{4} \times 1,144/2,300$, which is 286/2,300. The odds in favor of the queen's falling on the lead of ace and king are therefore 1,222 to 1,078, or 1.13 to 1. Hence the rule to lead ace and king if the queen and three are in the opposing hands.

We have already mentioned, in an earlier chapter, the fact that the number of possible bridge hands is 635,013,559,600. This means that if you write down on a slip of paper the names of thirteen cards, giving suit and denomination, the chance that your next bridge hand will contain exactly these cards is one in 635,013,559,600. The chance that your hand will contain thirteen cards of the same suit is one in 158,753,389,900. These facts, however, are not of great interest to you as a bridge player. They tell you only how exceedingly rare it is, if the cards are thoroughly shuffled, to hold the same hand twice, or to hold thirteen cards of one suit. It is of far greater interest to know the chance that your hand will contain a given distribution of length of suits. What, for example, is the probability that your next hand will contain a six-card suit, a four-card suit, a doubleton, and a singleton? We shall answer this question shortly, but first it is necessary to consider a possible difficulty.

We have just qualified a statement concerning the probability of dealing a given bridge hand by the phrase "if the cards are thoroughly shuffled." In our discussion of ten-spot and other spot-high hands in bridge, and in all our calculations in Chapters X and XI concerning the probabilities of poker hands, this same qualification was tacitly present. For all these calculations are applicable to the actual games only if shuffling is so thorough that the distribution of cards before the deal can be assumed to be *random*. We are entitled to ask, as a theoretical matter, what we mean by the word *random*, and we are entitled to ask, as a practical matter, how we can be sure that the shuffling is adequate to justify the assumption of randomness.

It is not difficult to illustrate what we mean when we say that the order of the cards is random. We need only be on our guard against reasoning in the following manner: There are no "privileged" orders in a deck of cards, for each of the possible orders is exactly as probable in advance as any other. Therefore the deck of cards is always in a random order. The fallacy in this reasoning lies in regarding the idea of randomness as static instead of dynamic. The order of a deck of cards is neither random nor nonrandom in itself. These ideas apply only when the order of the cards is being changed by a process of some sort, such as shuffling. And by imagining a particular sort of shuffling it is easy to illustrate what we mean by the word *random*.

Suppose that we think of the cards as numbered from 1 to 52, for example, by being placed in fifty-two numbered envelopes. It does not matter in what order the cards are placed in the envelopes. We imagine further that we are equipped with a special roulette wheel bearing the numbers from 1 to 52. Our shuffling will consist in spinning the roulette wheel fifty-two times and in placing the cards in the order indicated by the results. If this process were repeated before each deal of the cards, we could be certain that the order of the cards was random. Of course, any such method is so clumsy and tedious as to be out of the question in practice. We have imagined it only to make clear the meaning of the word *random*.

If the order of the cards is random, each deal is strictly independent of preceding deals, and the results of our calculations are strictly applicable. In the opposite case our results are not strictly applicable, and it is an impossible practical problem to determine to what extent, if any, they can be correctly used.

These questions concerning the adequacy of the shuffle are most interesting. It is perhaps unnecessary to point out that the problem is a purely practical one and can be settled only by actual observation of the fall of the cards in play. We encountered an analogous problem in Chapter VI in connection with the game of roulette. The question was whether or not the roulette wheel in practice follows the laws of chance, and we quoted some results obtained by Professor Pearson which indicate that there may be significant variations even in the case of the finest mechanical equipment. The question of whether a given phenomenon, such as the deal in poker or bridge, or the spin of the roulette wheel, follows the laws of chance is in reality a statistical problem, and more will be said on the subject in succeeding chapters devoted to statistics.

In the meantime we may remark that it is exceedingly difficult to determine, by casual observation of the fall of the cards, whether or not bridge hands follow the chance laws of distribution. The difficulty is not due solely to the immense number of deals that must be observed, but also to the fact that the laws of chance themselves prescribe what appear to be freakish runs of hands.

It would be a matter of considerable interest in the theory of bridge to submit to statistical analysis the records of a long series of actual hands, particularly if dealt by players who make a habit of reasonable care in shuffling. From such a record it could easily be determined whether or not shuffling was adequate to secure a random order. In the absence of a conclusive test, and with a word of warning as to the practical application of the results, we shall consider the mathematical theory of the distribution of cards in bridge hands.

Table XVII on page 187 gives the theoretical frequency of occurrence (based on a series of ten thousand hands) of the suit distributions indicated in the first and fourth columns. For example, the distribution indicated in the second line of the first column is 4432. This means that the hand contains two four-card suits, one three-card suit, and one doubleton. It does not matter, for the purposes of the table, which order the suits are in. Line two of the third column tells us that the distribution 4432 will occur, in the long run, 2,155 times out of ten thousand hands. This means that the probability that your next hand (or any other) will have the distribution 4432 is 0.2155, or a little better than 1 chance in 5.

We have given the frequencies to the nearest whole number, with the result that the distributions 8500 and 9400 carry the frequency 0. The actual values are respectively $\frac{3}{10}$ and $\frac{1}{10}$. Similarly, the expected frequencies of those distributions that contain ten or more card suits are too small to be listed. For example, the theoretical frequency of the distribution $\overline{10}111$ is 4/1,000 in 10,000 hands; that of $\overline{10}210$ is 1/100 in the same number of hands.

Owing to the fact that the number of suits and the number of hands dealt are both equal to four, this table can be interpreted in another way. It tells us also the probability that a given suit will be distributed among the four hands in the manner indicated. Interpreted in this way, the second line tells us that the probability that on the next deal any suit named in advance, say spades, will be divided among the players according to the scheme 4432, is 0.2155.

The mathematical theory of computing such a table is entirely analogous to that of the table of poker hands, which we discussed in some detail in Chapter X. We shall therefore not

TABLE XVII

Distribution of Bridge Hands

Distribution of Suits	Longest Suit	Frequency in 10,000 Hands	Distribution of Suits	Longest Suit	Frequency in 10,000 Hands
4441		299	7600		1
4432		2,155	7510		11
4333		1,054	7420		36
			7411		39
	4	3,508	7330		27
			7321		188
5530		89	7222		51
5521		317			
5440		124		7	353
5431		1,293			
5422		1,058	8500		0
5332		1,551	8410		5
			8320		11
	5	4,432	8311		12
			8221		19
6610		7			
6520		65		8	47
6511		71			
6430		133	9400		0
6421		470	9310		1
6331		345	9220		1
6322		565	9211		2
	6	1,656		9	4

delay our discussion of the table itself by repeating the theory here.*

* The reader who wishes to verify this table for himself may be interested in seeing the expression that leads to the probability in a sample case. The probability of the distribution 4432 is given by

$$\frac{C_4^{13} \times C_4^{13} \times C_3^{13} \times C_2^{13} \times 12}{C_{13}^{52}}.$$

The meaning of the symbol C_n^m is given in Chapter X. The last term in the numerator of the above expression, namely 12, represents the number of distinct orders of the four suits in the distribution 4432.

Table XVII gives us a good deal of information about probable distributions. In order to avoid confusion we shall interpret the table, in what follows, as applying to the distribution of suits in a single hand. We notice that there are exactly three types of distribution in which the longest suit contains four cards. The frequency of hands with suits of at most four cards is the sum of these three frequencies, as is indicated in the table. This sum is 3,508 out of 10,000, or a probability of approximately 0.35. The probability of holding a hand with the longest suit five cards is 0.44, which is the largest of all the corresponding figures. The odds against the longest suit's being four cards are 65 to 35, or almost 2 to 1.

The probability that your next hand will contain one or more singletons is 0.31, or a little less than 1 in 3. The probability that it will contain one or more blank suits is 0.05, or 1 in 20. These results are obtained by adding those probabilities in the table that correspond to hands with singletons, or blanks, as the case may be. In the same way, other probabilities of some interest can be obtained from the table. Although it is more probable that the *longest* suit in your next hand will be five cards than any other number, this of course does not mean that a five-card suit is more probable than a four-card suit. The probabilities, taken from the table, of holding one or more suits of the length indicated, are as follows:

TABLE XVIII

One or More	Probability
Four-card suits	0.67
Five-card suits	0.46
Six-card suits	0.17
Seven-card suits	0.035
Eight-card suits	0.005

These probabilities add up to more than unity. The reason for this is that there are distributions that include more than one of the above classifications. For example, the distribution

TABLE XIX

DISTRIBUTION OF BRIDGE HANDS

ASSUMING THAT ONE HAND CONTAINS THIRTEEN CARDS OF THE SAME SUIT

Distribution of Suits	Longest Suit	Frequency in 10,000 Hands
553		1,750
544		2,430
	5	4,180
661		141
652		1,273
643		2,592
	6	4,006
760		22
751		212
742		707
733		518
	7	1,459
850		12
841		88
832		212
	8	312
940		4
931		20
922		16
	9	40
$\overline{10}$30		1
$\overline{10}$21		2
	10	3

5431 includes both a five-card suit and a four-card suit. The odds, then, are 2 to 1 in favor of holding at least one four-card suit, and about 11 to 9 against holding at least one five-card suit.

Table XVII has told us the probability of the various possible distributions in a player's hand. Suppose, however, that the player has already picked up his hand and that it contains a certain distribution, say six spades, four clubs, three hearts, and no diamonds. The problem of interest then is to determine the probable distribution of the remaining thirty-nine cards in the three other hands. There are seven spades, nine clubs, ten hearts, and thirteen diamonds. Table XVII gives him no information whatever on this point. The problem can be solved by the same methods used in constructing Table XVII, but in this instance the calculations are very much more laborious, due to the fact that suits must now be taken account of, and to the fact that new calculations must be made for each of the possible distributions in the player's hand. We shall content ourselves by indicating the probabilities of the possible distributions in the other three hands for the extreme case where the player who has examined his hand holds thirteen cards of one suit. This particular case is of slight practical value, but it will serve to indicate how these various probabilities change, and to set certain limits on them. The result of these calculations is contained in Table XIX.

Table XX gives a comparison in summary form of Tables XVII and XIX, using probabilities in place of the corresponding frequencies per ten thousand hands.

In interpreting Table XX (and of course Table XIX), we must keep in mind that the probabilities in the last column are based on the most extreme unbalanced distribution that is possible. In actual practice we may therefore expect the probabilities of distributions in the other three hands to fall somewhere between the values given in column two and those in column three. If your hand is of the balanced-distribution type, the expected probabilities will be closer to those of column two.

It is interesting in this connection to notice that the probabilities in the case of hands with the longest suit five cards differ little in value. This means that regardless of the distribution of suits in your hand, the probability that your partner's hand, for instance, contains a five-card suit, but none of

TABLE XX

DISTRIBUTION OF BRIDGE HANDS

Longest Suit in Hand	Probability on Deal	Probability if One Hand Holds Thirteen of One Suit
4	0.3508	0
5	0.4432	0.4180
6	0.1656	0.4006
7	0.0353	0.1459
8	0.0047	0.0312
9	0.0004	0.0040
10	0.0000	0.0003

greater length, is between 0.42 and 0.44. The most significant change in the probabilities in moving from column two to column three, outside of the obvious impossibility of a hand with no suit longer than four in the latter case, is the very large increase in the probability of six-card suits. In this case it is to be noted that the probability fluctuates widely according to the distribution in your hand.

If we go one step further than we went in Table XIX and assume that the player has seen both his own hand and dummy's, we are back to distribution problems of the type that we considered earlier in this chapter. There we were interested in finding the probability that the Q will fall on the lead of A and K, if the Q and three are in the two unseen hands.

TABLE XXI

PROBABILITY OF HONORS

	Probability of	
	One or More	Two or More
Four-Card Suit Honors A	$\dfrac{220}{715}$	. . . .
AK	$\dfrac{385}{715}$	$\dfrac{55}{715}$
AKQ	$\dfrac{505}{715}$	$\dfrac{145}{715}$
AKQJ	$\dfrac{589}{715}$	$\dfrac{253}{715}$
AKQJ10	$\dfrac{645}{715}$	$\dfrac{365}{715}$
Five-Card Suit Honors A	$\dfrac{495}{1,287}$	. . . .
AK	$\dfrac{825}{1,287}$	$\dfrac{165}{1,287}$
AKQ	$\dfrac{1,035}{1,287}$	$\dfrac{405}{1,287}$
AKQJ	$\dfrac{1,161}{1,287}$	$\dfrac{657}{1,287}$
AKQJ10	$\dfrac{1,231}{1,287}$	$\dfrac{881}{1,287}$

We can come closer to the mathematical theory that lies behind a system of bidding by taking account of the probability of holding various combinations of honor cards in suits of various lengths. The probability that a four-card suit will contain an A (or any other card named in advance) is simply $\frac{4}{13}$. For a suit of any length, say n, the corresponding probability is simply $n/13$. The probability that a four-card suit will contain both the A and K is found by dividing the number of distinct ways in which two cards can be selected from eleven cards (the number of cards left after picking out the A and K) by the total number of ways that four cards can be selected from the thirteen. The result is 55/715, which equals 11/143.

The calculations in the more complicated cases are entirely similar. Table XXI on page 192 is computed in this manner. It gives the probability, once you know that a suit contains four cards, or five cards, that it will include one or more, or two or more, of the combinations of honors indicated in the left-hand column. It is to be remembered that it makes no difference in these calculations which particular honors we use for the purpose; for example, the table tells us that the probability of holding two or more of AKQ in a four-card suit is 145/715. Our statement means that the same probability applies to any other choice of three honors, such as AQJ, or KJ10. It would be easy to extend this table to suits of any length.

By combining Tables XVII and XXI, we can solve such problems as the following: What is the probability that your next hand will have at least one four-card suit containing two or more of AKQ? The manner of accomplishing this will be made clear by Table XXII below.

The probabilities in the second column are obtained from Table XVII. The first probability in the third column is obtained from Table XXI, translated into decimals. It tells us that the probability that *one* four-card suit will contain two or more of AKQ is 0.203. It was necessary to deduce from this value the corresponding probability in the case of *two* four-card suits, and likewise for *three* four-card suits. This is equivalent to a problem that we solved in Chapter V in connection with throwing dice. The probability of throwing a six spot with one die is $\frac{1}{6}$. What is the probability of throwing at least one six spot with two dice? We find the answer, which is $\frac{11}{36}$, by asking first the question: What is the probability of throwing *no* sixes with two dice? And this probability subtracted from unity (or certainty) gives us our result.

TABLE XXII

Number of Four-Card Suits	Probability on Deal	Probability of Two or More of AKQ	Product of Columns Two and Three
1	0.409	0.203	0.083
2	0.228	0.365	0.083
3	0.030	0.494	0.015
		Required probability	0.181

In obtaining the other two numbers in column three we have followed a similar procedure. When we know the probability that your next hand will have, for example, exactly two four-card suits, and when we know the probability that at least one of these suits will contain two or more of AKQ, the total probability is obtained by multiplying, as is indicated in the fourth column. Upon adding the numbers in column four we obtain the probability that we set out to compute. This gives us the chance that your next hand will contain at least one four-card suit with two or more of AKQ. The result is 0.181, or a little less than 1 chance in 5.

Let us carry through the similar calculations for the longer suits. We then obtain the following:

TABLE XXIII

Number of Five-Card Suits	Probability on Deal	Probability of Two or More of AKQ	Product of Columns Two and Three
1	0.417	0.315	0.131
2	0.041	0.531	0.022
			0.153
Number of Six-Card Suits			
1	0.165	0.437	0.072
2	0.001	0.683	0.001
			0.073
Number of Seven-Card Suits			
1	0.035	0.563	0.020
Number of Eight-Card Suits			
1	0.005	0.685	0.003

We can summarize these results as follows:

Length of Suit	Probability
4	0.181
5	0.153
6	0.073
7	0.020
8	0.003
	0.430

For suits longer than eight cards the probabilities are negligible.

The sum of all the probabilities is 0.430, as indicated. We have now arrived at the following conclusion: The probability that your next hand, or any other random hand, will contain at least two of AKQ in some suit longer than three cards is 0.43 In other words, the odds are 57 to 43 against such a hand on the deal. This result, as computed, includes the probability that you will hold two or more of AKQ in two or more suits.

The reasoning that we have used to obtain the result just given is not entirely correct. It is a short cut that leads to a result very close to the accurate one, resembling, in certain respects, the approximate method that we adopted in Chapter XI in discussing poker strategy. In the present instance we are not justified in adding the probabilities obtained separately for suits of various lengths. For in doing so we have assumed that the probability of the given honor holding in a distribution like 5431 is the sum of the separate probabilities for four-card and five-card suits. But this probability must be computed according to the same principle that we used in a distribution containing two suits of equal length, such as 4432. When the full computation is carried out according to correct principles, the result obtained is 0.404. Thus our approximate method has overstated the probability by 0.026, which is too small an error to have any practical bearing on the problem.*

We have given the probabilities for only one case, namely for hands that hold two or more of AKQ. Under this case we have listed the probabilities separately for suits of from four to

* The manner in which the accurate calculation is carried out is as follows: Consider, for example, the distribution 5431. The probability that either the four-card suit or the five-card suit, or both, will contain at least two of AKQ, is 0.203 + 0.315 − 0.203 × 0.315, or 0.454. This is to be multiplied by the probability of the distribution 5431, from Table XVII, which is 0.1293. This gives the value 0.0587. Each line of Table XVII is to be treated in similar manner, and the results added. The following examples of various types will make clear the process:

Distribution	Probability of Distribution	Probability of 2 or More of AKQ	Product of Columns Two and Three
4333	0.1054	0.203	0.021
4432	0.2155	0.365	0.079
4441	0.0299	0.494	0.015
5422	0.0158	0.454	0.048
5440	0.0124	0.565	0.007

eight cards, and we have combined these separate probabilities into one that represents the probability for any suit of four or more cards. We have selected this particular case because it provides so good an illustration of the application of the laws of chance to a wide range of similar problems, and not at all because it is of outstanding significance in the game.

If, on the other hand, similar tables were constructed by the same method for other combinations of honors (and other cards) that are of significance in the bidding, it is clear that they would provide the basis for a mathematical approach to the problem of rating the strength of hands. The results would have a certain analogy to those obtained in Chapter XI concerning the average "go-in" hand in poker. But to carry through to the point of building a full mathematical approach to the problem of correct initial bidding, to the formulation of a system of bidding, in other words, it is necessary to go much further. It is necessary to compute also the probabilities of various honor holdings for the other three hands, *after* one player has examined his hand. We have taken the first step in this direction in Table XIX for the extreme case, in which the calculations are much simplified, where the first player holds thirteen cards of one suit. It is possible, although only at the expense of a great deal of labor, to do the same thing for each of the distributions listed in Table XVII. Then for each combination of honors that seems significant the probabilities can be computed, just as we have done for the special case that we selected as an illustration.

In the first edition of this book (1939) we said, referring to the long calculations just outlined: "It is unlikely that anyone will care to undertake these computations, but it would be interesting to compare the indications given by such a study with the bidding systems that have been arrived at by empirical and intuitive methods." This has since become possible. In fact, as the above lines were being written it is probable that a part, at least, of these calculations had already been made. For the following year there appeared in France a very interesting book* on the mathematical theory of bridge. The subject is developed along the lines of the present chapter, but the authors, using electric calculators, have carried the work through to the point of giving complete tables of the probabilities of the various distributions, corresponding to our Tables XVII and XIX, for each stage of the game, before the deal, after the deal, and after dummy has been laid down. These calculations can serve as a basis for the building of a solid system of bidding.

* F. E. J. Borel and André Chéron, *Théorie Mathématique du Bridge*, Gauthier-Villars, Paris, 1940.

PART TWO

Statistics

CHAPTER XIV

From Chance to Statistics

WE HAVE now completed our survey of the theory of probability and of some of its more immediate applications. For illustration we have drawn on a fairly extensive list of games of chance, where the laws of chance appear in their simplest form. As previously pointed out, the theory of probability grew out of simple gaming problems, and its early successes were in that field. Today, however, its successes are spread through almost every field of human endeavor, and its applications to simple games are chiefly of interest as an ideal introduction to an understanding of its principles and methods.

One of the most potent and fruitful applications of the theory of probability is in the field of statistics, to which the balance of this book will be devoted. We have already given a tentative definition of statistics as organized facts, and of the theory of statistics as the art of selecting and manipulating these facts in such manner as to draw from them whatever conclusions may soundly be drawn. When these facts are numerical, or can in any way be related to numbers, many of the powerful techniques of modern mathematics are directly applicable. Without these techniques it must be admitted that statistics would be a rather dreary business. For this reason it is the practice to confine the sets of facts studied to those that can be related to numbers, and we shall follow that rule here.

Formal definitions of complex ideas have a way of being elusive, incomplete, or even misleading. No one, for example, has ever given a really satisfying definition of mathematics. As statistics is in fact a mathematical science, it would be too much to expect that our definition is more than a mild indication of what we are talking about. One might ask, "What precisely do we mean by *organized* facts? How do we determine whether a set of facts is organized or not?" One might ask how to find out whether or not a set of facts can in some way be connected with numbers. And how? The most satisfactory answers to such questions are to be found in examples of statistical problems and their solutions. In this way the reader will first grasp the spirit of the methods used; then he will understand what sorts of facts are grist for the statistical mill, and why. One of our leading mathematicians has advised his students to read rapidly through a new field, even though they understand only the broad outlines, before getting down to the job of mastering it. Keeping oneself oriented in this manner is good procedure in any technical field.

The end result of a statistical study is a set of conclusions. These conclusions can usually be applied to the future and are therefore predictions of things to come. We noticed the same situation in connection with the theory of probability. To say that the probability of throwing a six with one die is $\frac{1}{6}$ is equivalent to saying that in a long series of *future* throws about one sixth will be sixes. In this form of prediction we do not attempt to specify each individual event; we merely list the various possibilities and attach to each a number giving its frequency of occurrence.

Suppose that we have before us a record of 12,000 throws of a die, and that by actual count we find that 1,980 were sixes. This is a statistical record on which we can base a statistical conclusion. Dividing 1,980 by 12,000, we find that the six came up approximately once in 6.06 throws. We are tempted to conclude that in a future series of throws the six will appear with about this same frequency, and this conclusion will be sound if we can satisfy ourselves that 12,000 throws is a large enough sample. This question of the adequacy of a sample is basic in statistics, and we shall discuss it in some detail later.

We have approached the problem of predicting the fall of dice from the point of view of both probability and statistics and have arrived at approximately the same conclusion. Probability said, The six will appear about once in six throws. Statistics said, The six will appear about once in 6.06 throws. The difference between the two answers is small. If it were not, we should have to conclude either that the statistical sample was too small or that the die was not symmetrical.

This simple example shows how closely related are the theories of probability and of statistics. We shall see this relationship even more clearly as we proceed.

The science of statistics, like a manufacturing process, begins with raw materials, which are the crude data, and through the refining processes that follow turns out the finished product, which should consist of whatever valuable conclusions lay buried in the data. Just as the manufacturer, before building his factory and installing his equipment, must specify in detail each quality of the finished product, so in statistics the first basic element is to know precisely what is to be investigated, to formulate the problem as accurately as possible. For the quality of the conclusions depends in part on the exactness of this formulation. It depends in part on the appropriateness and accuracy of the data, and in part on the technical skill and "horse sense" that go into the handling of these data.

Before we say anything further about the methods of statistics, and of their relation to the theory of probability, it will be well to have clearly in mind the nature of some of the important types of statistical problems. Here, therefore, is a random list of a few typical problems, including many that will be discussed in this and later chapters. For convenient reference we have numbered the problems, and to avoid headings and repetitions we have first described the data; then, in each instance, we have given one or more possible fields of application for a statistical investigation using these data:

1. Height of American males.
   In tailoring or in coffin manufacturing.
2. Roulette records.
   To determine whether the apparatus has a bias; in other words, to determine whether the laws of chance apply.
3. Fire from mechanically held rifle.
   To find ways to improve various ballistic features.
4. Railroad passenger traffic.
   To determine in advance the most probable service requirements.
5. Sale of items in a retail store or mail-order house.
   To estimate future demand.
6. Birth and death records of a group of the population.
   The business application of these data is to life insurance. They are also of value in sociology.
7. Records of defective manufactured parts and corresponding cost figures.
   To determine the most advantageous point in the manufacturing process at which to reject defective parts.
8. Inherited characteristics.
   To verify and extend the Mendelian theory of heredity.
9. Crop production and prices.
   For government to determine national farm policies.
10. Number of kernels on an ear of corn.
    To find methods of maximum production.
11. Statistics on incidence of various diseases.
    To formulate public health policies.
12. Precision measurements.
    To determine from a set of measurements which value it is most advantageous to use, and to find some measure of the reliability of this value.
13. Data on the length of service of employees.
    To set up a system of pensions.
14. Mining accidents.
    To reduce risks in mining.
15. Responses to advertising.
    To determine the relative values to the advertiser of two or more advertising mediums.

16. Results of a given treatment in a given disease.
    To determine the effectiveness of the treatment.
17. Heights of fathers and sons.
    To find the relation, if one exists, between the heights of fathers and sons.
18. Income and number of offspring.
    To determine the relation, if there is one, between family income and number of offspring.
19. Sightings of enemy submarines.
    To determine the probable position of the submarine, with a view to sinking it.

These examples illustrate several distinct types of statistical problems, each of which merits careful discussion. One of the most fundamental problems of statistics is to determine whether or not a given sample is adequate to represent an entire population. In statistics the word *population* is used in an extended sense to mean any class of things, whether composed of people or not, such as all coal mines in West Virginia. In many of the problems listed as examples the first essential is this determination of the adequacy of the sample. Back in the days of the W.P.A., Surgeon General Parran made a survey of the health records of 2,660,000 individuals in order to determine the effect of illness on our economy. This survey included people from every part of the country, from city, town, and farm, from every economic level, and from all age groups. Assuming that his sample fairly represented the total population, then about 130,000,000, the Surgeon General drew certain conclusions, of which an example is: "Every man, woman and child (on the average) in the nation suffers ten days of incapacity annually." The accuracy of such conclusions is a matter of great importance, and it depends above all on the adequacy of the sample of just over 2 per cent of the population of the country. It is evident from this quotation that an effort was made to select these 2,660,000 individuals in such manner as to mirror as closely as possible the population as a whole, at least as regards those qualities with which the investigation is occupied. In order to accomplish this, it is necessary to take account of many more factors than those mentioned in this quotation, and no doubt many more were actually considered. For each factor that is taken account of, it is necessary to include in the sample in proper proportion the corresponding class of individuals. (A class of the population, statistically considered, is merely a set of individuals having some definite quality in common, such as living in California, or being between fifty and fifty-five years of age, or being unemployed.) Since the number of separate statistical classes that made up the total sample of 2,660,000 persons is large, it follows that the number of individuals in each of these classes must have been *relatively* small. Clearly, too, the smaller the number of individuals in a sample, the less reliable the sample. This is a point to which we shall have occasion to return more than once.

Reaching conclusions about an entire population from the statistics of a sample is like enlarging a photograph. Small defects are magnified into serious blemishes. If a sample does not take proper account of the classes that make up the population it is supposed to represent, it is called a biased sample. This survey of 2,660,000 individuals, for example, would be a biased sample if no account had been taken of age, or of sex.

But the failure of a sample to take account of all possible distinctions by no means proves that it is a biased sample. It is only the omission of *essential* distinctions that renders a sample inadequate, and it is not always evident in advance which characteristics will turn out to be essential. This indicates a fundamental difficulty against which the statistician (and the critic of statistics) must be continually on guard. In a sense he cannot know with certainty how to collect the data for his sample until his investigation is in an advanced stage. He may find, for instance, after exercising every precaution, that his data contain some unknown disturbing factor, and this may be due to the failure to separate in his sample two classes that should have been separated. When the statistician is himself in control of the collection of the data for his investigation, the difficulty is not insuperable, although it may mean a large added amount of labor. But when he is forced to use data collected by others, perhaps for entirely different purposes, and possibly years ago, he is in danger of finding himself with a hopeless statistical problem on his hands. All this requires the soundest sort of judgment, for which no amount of technical mathematical skill is an adequate substitute. On the other hand, the statistician with a defective knowledge of higher mathematics will find that the technical aspects of the subject absorb an undue proportion of his attention and energies, to the detriment of his handling of the investigation as a whole. If he has no knowledge of higher mathematics, he is incompetent and has no business to conduct a statistical investigation.

In addition to the danger of reaching false conclusions due to a biased sample, there is the danger of reaching equally false conclusions because the sample is too small. A problem of the same sort from a game of chance would be this: If you were tossing pennies with a suspicious-looking stranger and every toss so far had resulted in heads, how many such tosses would you need in order to be convinced that the stranger was tossing a trick coin with a head on both sides? And how many tosses would you need if the stranger was *not* suspicious looking? The question of whether a given sample is sufficiently large is not only an interesting one, but one of great practical importance in statistics, and more will be said regarding it in the next chapter.

Another type of problem, illustrated in number 2 of the list on page 204, is to determine by a study of statistical records whether the phenomenon in question operates in accordance with the laws of chance. In the case of roulette, the least bias or defect in the roulette wheel, or dishonesty on the part of the croupier, is sufficient to throw the statistics off from those anticipated by theory. In our chapter on roulette we referred to a statistical study of just this sort which was made in 1894 by Karl Pearson, based on the record of 24,000 spins of the wheel. The conclusion was that in certain respects the roulette wheel did not follow the laws of chance.

In all such investigations it is important to keep in mind that the actual and the theoretical tables must agree, statistically speaking, in *every* respect. It is not enough to show merely that there is agreement on one or two major features, such as the fact that red and black have appeared just as often as the theory predicts. We must also show that the number of times that red appeared, say four times or nine times in succession, is in accord with the theory. On the other hand, if it can be shown that the ideal and the actual distributions *disagree* in *any* respect, we immediately compute the probability that such a disagreement would happen in the number of games in question. If this probability is extremely small, we shall be forced to believe that for some reason the roulette wheel is not operating according to the laws of chance. We can then confirm this belief most effectively by considering other features of the comparison, or by continuing the experiments, if that is possible.

Another problem of the same sort was discussed in connection with the game of bridge, back in Chapter XIII. The question there was whether or not the ordinary type of shuffling effectively produces a random order in a deck of cards. If it does, then the theoretical probabilities of various distributions on the deal are applicable. But if it fails to produce a random order, then there is a disturbing element present in the fall of the cards on the deal, and calculations based upon pure chance will not apply.

This matter of determining whether or not a given phenomenon follows the laws of chance has become one of extreme importance in many of the sciences. Many years ago decisive experiments in physics demonstrated that changes in the energy of the atom, which occur when the atom gives out radiant energy in the form of a pulse of light or absorbs radiant energy, take place according to statistical law; in other words, according to chance. In radioactivity, for example, when alpha particles (helium atoms) are shot out of radium, the same is true. The number of particles shot out seems to depend only on the number of radium atoms present and follows a statistical law. In modern astronomy we find many examples of statistics at work. It is now known that the tremendous energy of the sun is sustained primarily by a remarkable closed series of six reactions between the atomic nuclei of the elements hydrogen, helium, carbon, nitrogen, and oxygen, carried on in the sun's interior where the temperature is as high as 20 million degrees Centigrade. These reactions are known as the carbon cycle, because of the important role that carbon plays; it is not consumed in the process but acts as a catalyst. The net result is to convert hydrogen into helium. Each reaction takes place according to statistical laws. Everywhere we find inconceivable numbers of elementary particles, protons, electrons, neutrons, and so on, rushing about at mad speeds in apparent disorder, but actually in some way or other responding to the slow and orderly dictates of statistical law. What a radical change this view of things represents in less than fifty years!

In biology also statistics has played an important role. In 1866 Mendel published his observations and conclusions after years of experimenting in the breeding of plants. He established, through the statistics he had gathered, that hereditary characteristics are transmitted according to the laws of chance. This work led to the founding of the sciences of genetics and eugenics, with important applications to the breeding of plants and animals, and new possibilities for the improvement of the human race. Some years later Karl Pearson, whom we have mentioned in connection with his experiments on roulette, was the center of a distinguished group of British statisticians who vigorously applied statistical methods to the problems of organic evolution, heredity, and related fields, making important contributions to mathematical statistics itself as they did so.

This work has been carried on by Sir Ronald A. Fisher and his associates.

Our next illustration, number 6 on our list, has to do with life insurance, with which most of us are well acquainted. When a life insurance company fixes the amount of your annual premium, it is using a table of statistics, the mortality table, to determine your chance of survival from year to year. The form of life insurance in which the relationship between your premium and this table of statistics is clearest is what is known as term insurance. In this type you are buying only protection for your beneficiary, in case of your death during the year, while in other types the insurance company fulfills also the function of a savings bank.

Suppose that the mortality table indicates that at your age there is 1 chance in 50 of your death during the next year. Then an amount of your premium equal to 2 per cent of the face value of your policy is what we can call the *pure chance component*. The balance of your premium constitutes an amount sufficient to cover the operating expenses of the insurance company, and profits, if the company is not of the mutual type. It includes a safety factor as well, and this may amount to one quarter or one third of the annual premium. Each year, after the first, an amount corresponding to this safety factor is credited to your account with the company, usually in the form of dividends, so that the amount remains constant from year to year, unless current statistical experience requires that it be adjusted. A safety factor of some sort is obviously essential to the financial soundness of the insurance company. It covers not only possible inaccuracies in the original mortality table, but other contingencies such as changes in the death rate since the mortality table was compiled and adverse runs of luck, possibly due to an epidemic in which an unexpectedly large number of deaths occur. Such runs of luck are not at all contrary to the laws of chance but are predicted by them.

Still another type of statistical problem has to do with relationships between sets of statistical data. An illustration is number 18 in our list. Suppose that an investigator goes from house to house, perhaps in taking the census, and gathers data on the amount of family income and the number of children in the family. Corresponding to each income, say $3,000, there will be a certain number of families with no children, a certain number with one child, a certain number with two children, and so on. If there are fifty classifications of income, there will be fifty such lists of number of offspring. Or the data can be listed on a different plan. Corresponding to each number of offspring, say five, there will be a certain number of families with $3,000 income, a certain number with $3,100, and so on. The problem is: Does there exist a relationship of some sort between the number of children in the family and the family income? For example, is it true that, by and large, the smaller the income, the larger the family?

If we examine the data as analyzed above, and if we find that families with no children have, on the average, a 50 per cent greater income than those with five children, and if we find also that the incomes of families with from one to four children are intermediate between the two, and decrease with an increasing number of children, we shall consider this result as tending to confirm the theory that such a relationship exists. We shall not jump to a final and certain conclusion in the matter, because statistical conclusions depend on probabilities and therefore carry the same uncertainties. If you take a coin from your pocket, toss it five times, and get five heads, you will not immediately conclude that the coin is biased. You will know that in the long run you will obtain this result once in every thirty-two trials.

When there is a statistical relationship between two things, such as the family income and the number of children, or the heights of fathers and their sons, we say there is a *correlation* between the things. Of what does this correlation basically consist? Well, there are three conceivable ways in which things may exist with reference to each other. First, they may have no relationship at all. Obviously, we should be a strange-looking race of people if heights and weights were entirely unrelated. Second, the relationship between two things may be hard and fast—the greater the cubic content of a lump of lead, the more it weighs, to give an obvious illustration. If this were the relationship between the physical human characteristics of height and weight, then it would follow that a change of height necessarily corresponded to a change of weight. But there is, third, the statistical relationship, in which the relationship is not hard and fast, but a statement of probabilities. Thus, greater height *probably* means greater weight in a human body. There is a correlation between them.

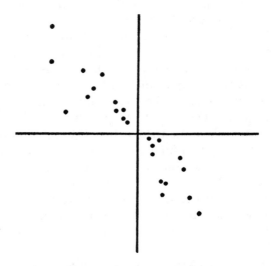

FIGURE I. SCATTER DIAGRAM.

We can express these three types of relationship in a visual way that is easily conceived. If, on an ordinary graph, we plot one of the two quantities that we are comparing horizontally, and the other vertically, we shall get three very different look-ing results, according to which of the three types of relation-ship holds. If there is a hard and fast relation between the two quantities, and to each value of one there corresponds one and only one value of the other, the chart will contain a single straight or curved line. If there is a statistical relation, the dots will fall into a sort of band, which may be either straight or curved. Such a configuration is called a "scatter diagram," and an example is given in Figure I. If there is no relation whatever between the two quantities, the dots will be scattered about the diagram in a random fashion.

It may be in order to point out at once that because two quantities show a high degree of correlation this does not neces-sarily mean that there is more than a purely accidental or an indirect relationship between them. For example, if we had before us a table showing in one column the height of corn in Illinois for each day of spring and early summer, and, in an-other, the amount of money spent by Illinois people for vaca-tions, we would find that there was a fair degree of correlation between these two quantities, and a scatter diagram would give a well-defined band of dots. But it is clear that there is no direct relationship involved.

Problem 17 on our list also concerns correlation. In a study of 1,078 British fathers and sons, given by Karl Pearson, it is shown that there is a correlation between the heights of fathers and sons. The betting odds favor a tall father's having a tall son, and a short father's having a short son. These odds are fixed by the appropriate statistical investigation. A few of Pearson's results are as follows: The average height of fathers was 5 feet 7.7 inches, that of sons, 5 feet 8.7 inches. The most probable height of the son of a father 6 feet tall turned out to be 5 feet 11.7 inches, that of the son of a father 5 feet 6 inches tall, 5 feet 8.6 inches.

A logical extension of Problem 17 would be obtained by taking account of the height of the mother as well. A typical result would be: The most probable height of the son of a father 6 feet tall and a mother 5 feet 2 inches tall is 5 feet 9 inches. (This figure is necessarily fictitious, as I do not know of any such investigation.) The problem could be still further extended by including the heights of daughters.

Several of the problems on our list, such as numbers 3 and 12, have grown out of laboratory experiments of one sort or another, and the same is true of the majority of the applications of statistics to science. In these cases the original data are almost invariably collected with a view to the statistical problem. They are usually the result of carefully planned and skillfully executed scientific experiments, and when this is the case the statistician's first problem is solved for him.

It is not necessary to say more than a word here about that part of statistical method which coincides so closely with the general experimental method in science. In a typical controlled experiment, the object is to determine whether or not a variation in one factor produces an observable effect. An example is: Can cancer in mice be inherited? Two groups of mice, as similar as possible, except for the fact that one group is cancerous, are subjected to conditions as identical as possible. Statistical records are kept for each group over hundreds or even thousands of generations. These records constitute the statistician's data. It is then up to him to determine whether or not these data indicate that under these conditions cancer in mice can be inherited and, of equal importance, to attach to his conclusions an estimate of their reliability.

When a relation of a statistical character is established between two or more things, it is called a statistical law. As this concept is at the root of all statistical thinking, and as it differs in important ways from other concepts of laws in science, we shall need to examine some of the characteristics of the notion of statistical law.

One of the great classical ideas that led to the development of modern science, as we know it, is that of natural law. According to this conception, a given set of circumstances lead without fail to a definite result which is stated by the law. Newton's law of gravitation is of this type, as is Einstein's law of gravitation. Such natural laws as these predict an event, when they predict it at all, in the form of a certainty. It is not stated that *probably* there will be a total eclipse of the sun on February 26, 1979, visible in Canada. The eclipse is definitely predicted, and the exact time of day, down to the second, is specified well in advance. If it turns out that the eclipse takes place one or two seconds early or late, the discrepancy is attributed to the extreme mathematical difficulty of the problem, and to possible unknown factors, and not to the presence of a chance element.

On the other hand, as we have seen, there is a second and quite different manner of predicting future events that modern science has found it necessary to adopt. For "the event will happen" this new type of law tells us "the chance that the event will happen is 1 in so many." In his famous prediction that a ray of light would be deflected by a heavy body such as the sun, Einstein was using the first of these forms. In predictions of other sorts of things, such as the next throw in a game of craps, the second of these forms is used.

In discussing how gases, like air, behave under certain conditions (when compressed, for instance) we should, strictly speaking, state our conclusions in a similar form, as probabilities. For gases are composed of enormous numbers of tiny particles moving at high speeds and continually colliding with one another, and the laws governing their behavior are *statistical laws,* which always lead to conclusions that require the adjective *probable.* As Eddington so well puts it, a pot of water on a fire "will boil because it is too improbable that it should do anything else." When the number of particles is extremely large, as in the case of the gas or the water, these conclusions may be so very probable that they need to be distinguished from certainties only when we are interested in the *why* of things, rather than the *how*. A cook who spent her time in pondering over the probability that her dishes would come to a boil would be more of a philosopher than a cook.

This first conception of natural law, which makes precise predictions, leads to the point of view called *determinism*. This doctrine states that the condition of the world at any one moment *determines* its condition at any future moment. This is not to imply that any living person has sufficient knowledge or penetration to make such a determination for more than a small, idealized fragment of the world; it is merely stated that all necessary data for the purpose are contained in a complete survey of the world at any instant.

This doctrine of determinism is quite distinct from that of *fatalism,* with which it is sometimes confused. The central idea of fatalism is that the events of this world are *preordained.* The fall of a drop of rain takes place not because of natural law but, in a sense, in spite of it. Fate has ordained that the drop should fall and has specified exactly how. In neither of these doctrines when pushed to their logical limits, is there a place for human choice, or free will.

In discussing many types of problems, for example how helium atoms are shot out from radium, science has had to leave room for the element of chance. The events of tomorrow are not quite determined by those of today. Certainties are replaced by a number of possibilities, only one of which will become the fact of tomorrow.

In the world of determinism, there is only one candidate running for each office; he is certain to be elected, and if we can find out his name and what office he is running for, our knowledge is complete. In the world of chance there are several candidates for each office, and we need to know also their individual prospects of election.

It is natural to inquire how it is possible for these two views of the world to live happily alongside each other. For certainly it is hard to believe that part of the world works one way, part another. And if all of it works the same way, which way does it work?

Until recent years physicists were almost unanimous in accepting the deterministic view as the correct one. They regarded the statistical approach as a valuable but artificial working technique, which provided a practical short cut in problems too complex for the human mind to spell out in full. Today, as we saw earlier in this chapter, statistics plays a major role in physics, and the majority of physicists believe that statistics, after all, was the invention of nature instead of man.

CHAPTER *XV* ·

Chance and Statistics

STATISTICS are tables of past facts. They may be compared in usefulness to the materials for a house that is about to be built. If the building materials consist of odd piles of stone, bricks, and lumber, the resulting house is likely to resemble a crazy quilt. It is only when the architect's plan is consulted, and when specifications and quantities for the material of each kind are laid out in advance, that a sane sort of house is constructed. The same is true of a statistical investigation. Unless the building materials, in the form of the data, are collected in the light of a unified plan, the resulting structure is not likely to have either stability or beauty.

The science of statistics thus consists of three indispensable parts: collecting the data, arranging them in the form of statistical tables, and, finally, utilizing and interpreting these tables. It is in the last of these steps that a mathematical technique is essential; frequently the data have been collected before the statistician begins his work. They may, for instance, cover a great many years. So, when the statistician does not collect his own data, his first step is to study the exact conditions under which the facts he has been given were gathered. If it is not possible to know these conditions accurately, the entire subsequent statistical structure will have to rest on an insecure foundation and be interpreted accordingly.

The science of statistics is permeated with the ideas of the theory of chance, from the first steps in collecting the data to the drafting of the final conclusions of an investigation, and in

218 ·

order to understand enough of this science to follow and to appreciate its widespread applications, it is necessary to look a little more closely than we have at the mechanism through which this union of ideas is accomplished. This we propose to do in the pages of this chapter. Since statistics is, after all, a technical subject, we shall not be able altogether to avoid certain topics that border on the technical. We shall discuss, for example, the methods that the statistician uses to study the play of chance forces that lie behind the statistical data before him, methods that draw on the theory of games of chance that we reviewed in the earlier parts of this book. It is only by doing so that we can grasp the true spirit of statistics and be prepared to trace, in succeeding chapters, its influence in many fields. The statistical machine has, furthermore, like many well-oiled mechanisms, a fascination of its own for those who take the trouble to watch it at work.

Crude facts are the raw materials of statistics; from them is evolved the finished product, usually a table of chances. A crude fact is a simple statement of a past event, such as "John Smith died yesterday at the age of forty." From large numbers of such crude facts the insurance actuaries, for example, have constructed mortality tables, which serve as a basis for computing the amount of your life insurance premium. This basic table gives the chance that a person of such and such an age will survive at least so many years. It tells us, among other things, that of one hundred babies born now, about seventy will be alive forty years from today. In other words, the chance that a child born now will reach the age of forty is approximately 7 in 10.

The choice of the crude facts that are admitted into the statistical table determines the scope and meaning of the resulting mortality table. If we compare the English, French, and American mortality tables, for example, we shall find no two of them alike. Each reflects the enormously complex factors, some of them peculiar to the country in question, and all of them beyond the powers of direct analysis, which determine the span of life.

If we include in the table only individuals who have been medically examined, and whose general health is determined to be above a specified level corresponding to the actual procedure in writing life insurance, the table will be significantly modified. If we admit into one statistical table only data concerning males, and into another only data concerning females, we shall find that the resulting mortality tables have important differences. If we construct mortality tables according to the occupations or professions of those whose vital statistics are included, we shall find, similarly, a number of distinct tables. If we adopted one of a large number of other classifications, there would result still other mortality tables. Which classifications are used in practice depends on the purpose for which the mortality tables are intended. The important thing is that the selection be made with this purpose in view. The corresponding thing is true not only in actuarial statistics, which is concerned with all forms of insurance, but in other statistical applications as well, whether the subject be biology or business.

Statistical tables are, then, the fruits of experience. They lead to the determination of probabilities which would otherwise be hopelessly inaccessible. We have seen, in the earlier parts of this book, that in many games of chance it is possible to compute the probabilities of the game entirely independently of all experience. After this has been done, these probabilities can be compared with the results of experience. To do this we used the "law of large numbers." We could also have started out by observing over long periods the fall of the cards or dice, constructing statistical tables, and finally obtaining from them the probabilities that we were looking for. In other words, we could have approached the subject from the statistical point of view.

In cases where it is possible to compute the probabilities in advance, as in simple games of chance, such an approach would be needlessly tedious and clumsy. But in more complicated matters experience, in the form of statistical tables, is the only possible road to a knowledge of the chances involved. In computing mortality tables, for instance, there can be no question of finding a priori probabilities. The causes that determine length of life are so complicated that we cannot even imagine what an analysis into "equally likely cases" would mean, although we see very clearly what they mean in throwing dice or dealing poker hands. In these complex situations we substitute *statistical probabilities,* determined by experience.

Not only are statistical tables the fruits of experience, but they are the fruits of large numbers of experiences. These experiences may consist of a great many repetitions of some event, as when one person throws dice a large number of times, or they may consist of the combined dice throwing of a large number of persons. In life insurance tables, the experience is necessarily of the second kind; for no one can die repeatedly, no matter how willing he is to sacrifice himself for the benefit of science, or of the insurance companies. In games of chance we have been able to predict with confidence how a player will come out *in the long run,* or how a *very large number* of players will come out in a single trial. Similarly in statistics, if our tables are based on a very large number of experiences, we can, in many cases, predict with confidence what will happen in the long run, whether we are concerned with an individual repeating an experience, or with a group of individuals.

This, perhaps, sounds too generalized, but its actual application to statistical problems is not difficult. Suppose that we have before us a table giving the heights of one hundred thousand American males, selected at random, and that we are interested only in learning something about the average height of Americans. It is simple enough, although a trifle tedious, to add all the heights together, and divide by their number, which is 100,000. This gives us the average height of the males in our sample. What have we learned by doing this? Can we assert that the average height of adult American males is very close to 5 feet 8 inches, if that is the result of averaging the sample? Before being able to make such an assertion with confidence, we should have to satisfy ourselves on several points.

We have been informed that our table of heights contains individuals "selected at random." If we are conscientious statisticians, or critics of statistics, we shall never pass this by without somehow assuring ourselves that the phrase "selected at random" means approximately the same thing to the person or persons who collected the facts for the table, as it does to us. We are faced with the question, raised in the preceding chapter, of determining whether a given sample is adequate to represent the entire population. Above all we should like to know on precisely what basis one man out of every four hundred was selected, and the other three hundred and ninety-nine omitted. With this information at hand, we shall soon be able to decide whether our idea of random selection corresponds to that used in making the table of heights. In the absence of precise information on this point, there are many questions to which we should require answers before being willing to attach weight to our conclusions: Does the sample correspond in geographical distribution to that of the population of the country? Does it include a representative distribution of ages? How many men of foreign birth are included? How many Negroes? Does the table show a preference for any particular class, social or occupational? On the other hand, if we knew that the selections were made by drawing names from boxes containing all the names, or in some equivalent way, we should feel satisfied that we are dealing with a fair sample.

Next, we ask whether a random sample of one hundred thousand is large enough for our purpose. We are attempting to reach a conclusion as to the adult male population of the United States, amounting to forty-odd millions of individuals, by using a sample containing measurements of only one hundred thousand, or 0.25 per cent of the whole. Before this question can be answered at all, it is necessary to know the accuracy that is required of us. If it is required to know merely whether the average height of an American falls between 5 feet and 5 feet 6 inches, or between 5 feet 6 inches and 6 feet, it is clear enough that a small sample is all that is needed. If it is required to determine this average height closely enough to feel confident that we know it to the nearest inch, a much larger sample is necessary, and so on.

Once given the degree of accuracy that is required of us, there remains the problem of determining whether or not the sample of 100,000 measurements is large enough to furnish it. This problem is typical of a whole class of statistical questions, to answer which a part of the mathematical theory of statistics has been built. We are not concerned here with the precise mathematical methods used to solve such problems; they belong to the professional kit of the statistician and are set forth in technical treatises. We are concerned rather to catch the spirit of the methods, to grasp the character and significance of the conclusions reached, and above all to understand the precautions that must be taken before accepting these conclusions.

We have before us the list of 100,000 measurements of heights, and we have also the average of these heights. In amateur statistics, the study of an experience table usually begins and ends with taking the average. While it is true that the ordinary arithmetical average is an important first step, it is but one of several essential characteristics of a table, and it is seldom true that the average alone is of value.

It is by considering also features of the table other than the simple arithmetical average that the statistician reaches his conclusions as to the adequacy of the sample of 100,000 heights. His next step is to compute a quantity called the *standard deviation*, which measures the extent to which the heights are scattered about their average, in other words their *dispersion*.

The amount of dispersion is evidently of major importance in the interpretation of the sample. We shall not stop here, however, to explain in detail how the standard deviation is computed; this will be done later in connection with a simple example, where the process can be easily illustrated. We shall merely remark that the standard deviation is itself an average.

In practice, when the statistician is confronted by a mass of data, such as the 100,000 heights, before making any computations whatever he arranges them in a frequency table. To do this he first chooses a limit of accuracy in listing the data, which he calls the *class interval*. Let us assume that in this case he selects half an inch. Then he enters opposite the height 68 inches, for example, the number of heights in the original data that lie between 67.75 inches and 68.25 inches. By making this table he shortens considerably the labor of computing the average; for he can first make a fairly accurate guess as to its value and then find out, by a mathematical process, how far off this guess is.

After entering each of the 100,000 heights in its appropriate place, he has before him what is known as a *frequency distribution*. This is the same kind of table as the one given on page 232. It tells at a glance how many men out of one hundred thousand in the sample are within $\frac{1}{4}$ inch, in either direction, of any one particular height, say 6 feet. It also tells at a glance the most popular or "stylish" height, called the *mode*, opposite to which the greatest number of individuals are entered.

Suppose, for example, that opposite 6 feet, or 72 inches, we find the figure 2,000. This means that $\frac{1}{50}$, or 2 per cent, of the individuals in the sample are 6 feet tall, to the nearest $\frac{1}{2}$ inch. Assuming that the sample adequately represents the entire population, we can express this result as follows: The probability that an American male *selected at random* is 6 feet tall to the nearest $\frac{1}{2}$ inch is 1 in 50. For each possible height other than 6 feet we can discover the corresponding probability in the same simple way. Thus our statistical table has led us to a table of chances.

What we wish to emphasize here is that the statistician's conclusion, when he reaches it, is in terms of chance or probability. He tells us the probability that the average value of 5 feet 8 inches is correct to within a certain amount, say $\frac{1}{10}$ of an inch. He may say, "The betting odds are even that the true value of the average for the entire population is within $\frac{1}{10}$ of an inch of that of your sample. It follows that the betting odds are $4\frac{1}{2}$ to 1 that it is within $\frac{1}{5}$ of an inch, and 100,000,000 to 1 that it is within $\frac{9}{10}$ of an inch." This is the sort of answer that we might have expected in the beginning and, in fact, the sort that should satisfy us the most.

When statistics answers a question for you, always look for a tag of some sort carrying a reference to chance. Its absence is a clear danger signal.

In order to see as clearly as possible the intimate relation that exists between statistics and the theory of chance, let us examine a situation of an entirely different sort, where we shall obtain a glimpse of a statistical law in action. When a rifle is fired at a target, the shots group themselves in a pattern that is more or less spread out; no matter how good a shot is behind the gun, there is always a certain amount of this spreading, or *dispersion*. After a few shots have been fired, the marks on the target are usually quite irregularly placed; there is no semblance of order.

If, however, many rounds are fired, there appears a definitely ordered pattern. There is a center about which the shots are densely packed in all directions; farther from it there are less and less shots, until finally a circle is reached outside of which there are none at all. (We are assuming that the rifleman is a good enough shot to hit the target every time, or, what comes to the same thing, that the target is large enough for him to hit.) This center may or may not coincide with the bull's-eye. When it does not, there is a *systematic* error of some sort, such, for example, as an error in the sights. On the other hand, the errors that cause the spreading about the center of the pattern are called *accidental;* they are due to one or more of a large number of causes, each small in itself: unsteadiness of hand, eccentricities of vision, variations of charge, irregular currents of air, and many others. The important thing about these accidental errors is their indifference to direction. Each one is as likely to be above as below the center of the pattern, to the right as to the left of it.

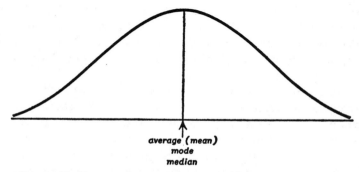

average *(mean)*
mode
median

FIGURE II. NORMAL FREQUENCY DISTRIBUTION (GAUSS'S LAW OF ERRORS).

In this illustration, which is typical of one class of those to which the statistical method can be applied, we see order emerging out of chaos. The more complete the underlying confusion, the more perfect the resulting order. Moreover, the ordering process is a progressive one; as more and more rounds are fired, it becomes more and more pronounced.

Precisely what is meant when it is said that the shot pattern is in accord with statistical law? Does it mean that the grouping of the shots about the center can be predicted in advance of the firing? Clearly not, for, apart from the systematic errors, nothing would then distinguish the expert shot from the man who has never had a rifle in his hands. Let us leave aside the systematic errors, which are utterly unpredictable and which can, for our purposes, be eliminated. Then the answer to our question is as follows: By use of the fact, mentioned above, that the accidental errors are indifferent to direction, and of one or two additional assumptions of a technical nature, it can be shown that the grouping of the shots is described by a statistical law known as Gauss's Law of Errors.

The mathematics involved in arriving at this result is beyond the scope of this book. It is important to understand, though, that such a law exists.

Gauss's law expresses the number of errors of each different size that will occur in the long run. It says that the smaller the error, the more frequently it will occur (the greater its probability, in other words) and that the probability of larger errors is progressively smaller. The curve drawn in Figure II above is an illustration of Gauss's law. Every law of this nature contains, in addition to one or more variable quantities, the familiar *x, y,* and so on, of algebra, a few undetermined numbers. These numbers vary from one problem or from one application to another, but are constants throughout each investigation. A little reflection will show that the presence of such numbers is essential. Something in the law must serve as a measure of those characters that distinguish each individual case. Something must distinguish between one measuring machine and another, between one observer and another, between one rifleman and another.

In Gauss's law there are four of these undetermined numbers. Two of them depend on the position on the target of the center of the pattern; the third and fourth depend on the amount of spreading, or dispersion, vertically and horizontally, and are a measure of the accuracy of rifle and rifleman. Again we come across the standard deviation! As we have already assumed that the center of the pattern coincides with the point aimed at, the first two of these numbers are in fact determined. There remain the third and fourth, which must be found statistically; that is, by actual firing at the target. If this were not so, if these latter numbers were determined by the statistical law, we should know at once that a mistake of some sort had been committed; for merely assuming the existence of accidental errors could not conceivably tell us anything about the marksmanship of the rifleman.

We have not as yet introduced the idea of statistical probability. To do so we can imagine the target to be divided into small squares, or, still simpler, into concentric rings about the center, in accordance with the usual practice in target shooting. In the latter case it is essential to assume that no systematic errors are present. Suppose now that a large number of rounds are fired, say 100,000, and that the number of shots in each of the rings is counted. The statistical probability of hitting any one of the rings is the number of shots in that ring, divided by the total number of shots, 100,000. Suppose that there are six rings (including the bull's eye) and that the number of shots in the one containing the bull's-eye is 50,000, the numbers for the others being successively 30,000, 10,000, 6,000, 3,000, 1,000. Then the corresponding probabilities are $\frac{1}{2}$, $\frac{3}{10}$, $\frac{1}{10}$, 6/100, 3/100, and 1/100. If the firing is continued, the betting odds are even that a given shot lands in the bull's-eye, 7 to 3 against its landing in the next ring, and similarly for the others. In practice, these probabilities will follow closely those indicated by Gauss's law.

Thus the statistical approach to target shooting has led us to a definite set of chances and so to a prediction of the future fall of the shots. But this prediction will hold good only as long as the gunner holds out; it is subject to all the frailties that his flesh is heir to. He may become fatigued, or get something in his eye, or one or more of a thousand other things may happen to spoil his aim. If we eliminate him altogether and imagine the rifle mechanically held in position, as is done in the munition plants, our results have to do with the quality

of arms and ammunition, instead of arms and the man, and
are less subject to sudden and unpredictable variations.

Let us compare briefly the important features of these two
statistical situations that we have used as illustrations: the
stature of American males, and the shot pattern on the target.
We shall find that they correspond rather closely. To the center
of the shot pattern corresponds the average height of American
males, as computed from the sample of one hundred thousand
individuals. To the scattering of the shots about the center,
which we called the dispersion, corresponds the spread of
heights about the average. Furthermore, we have seen that
the tiny accidental errors responsible for the spreading of the
shot pattern are as likely to deflect the shot upward as down-
ward, to the right as to the left, and that as a consequence
the grouping of the shots follows Gauss's Law of Errors, a form
of statistical law. In the case of the heights of American males,
we cannot speak of "accidental errors," for the phrase would
have no meaning. But it is nevertheless an *experimental* fact
that the grouping of the heights about their average closely
follows Gauss's Law of Errors.

When the numbers in a statistical table (arranged as a frequency distribution) follow the Law of Errors, it is called a
normal or *symmetrical* distribution. The general appearance
of such a distribution, illustrated in Figure II, shows that the
values are symmetrically placed about the vertical line that
represents the average of all the values, which explains why it
is referred to as symmetrical.

Thus the scattering of the shots and the distribution of the
heights both follow Gauss's law. But it is to be particularly
noticed that while, in the case of the rifle firing, it was possible
to say *in advance* of the actual firing that this law would be
followed, no such statement could be made in the case of the
measurements of heights. In the latter case, this fact was dis-

covered *after* the measurements had been made and entered in the table.

In a simple game of chance, as we have seen, it is possible to predict in advance of any trial or experiment not only what sort of distribution of possible cases will result—that is, what statistical law is followed—but the relative frequency of occurrence, or probability, of each of these cases. This is going one step further than it was possible to go in considering the target

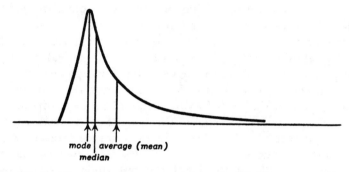

mode | average (mean)
median

FIGURE III. SKEW FREQUENCY DISTRIBUTION. The original data represent sales from a mail-order catalog. The curve represents the result of averaging and smoothing the data by the appropriate methods.

shooting. We can therefore arrange the three situations in order of increasing complexity as follows: simple game of chance, target shooting, table of heights.

We have previously encountered many examples of simple games of chance, and still another will be given shortly. As examples of the second of these classifications, we have all varieties of gunnery and artillery and the theory of measurements. Under the third classification come the great majority of statistical situations, in insurance, in the sciences, in business.

It would be a great mistake to conclude that, because the distribution of men's heights follows Gauss's law, this is usually the case in similar statistical studies. If we had weighed the hundred thousand individuals, instead of measuring their heights, we should have obtained a distribution that is *not* symmetrical. Such distributions are called *asymmetrical* or *skew* (see Figure III).

This fact about weights impresses most people as astonishing. "If heights form a symmetrical table or curve," they say, "why don't weights?" Their feeling is akin to that of the majority of mathematicians and statisticians for several decades following the death of Gauss. Because the exact assumptions required to produce Gauss's law had not yet been completely formulated, and because the authority of the great Gauss hung heavily on them, they tried their best to force all statistical series into the mold of the Gaussian Law of Errors. All true statistical series must be symmetrical, they argued, and concluded that asymmetrical series cannot represent *true* statistical data. They assigned various sources of the difficulty, that the data are inadequately analyzed, or that the number of cases is not large enough, for example. Such a situation does not exist in tossing coins or in throwing dice, where the distribution becomes more and more symmetrical the larger the number of trials. We shall see an example from dice throwing farther on. But isolated instances do not justify the conclusion that all statistical series are symmetrical. The overwhelming majority actually are not.

Let us try to rationalize the situation, at least partially. In obtaining the Law of Errors, which is symmetrical about the average, the assumption was made that the small elementary errors are as likely to be in one direction as in the opposite direction. Suppose that we were to assume, on the contrary, that each error is twice as likely to be in one of the directions, say north, as in the other, south. Then there would be a piling up of errors to the north. The curve would not be symmetrical. This example is sufficient to indicate that the symmetrical is a special case.

All varieties of skew distributions are met with in statistical practice. In a normal distribution (one that follows Gauss's law), more of the values in the table are packed closely about the average than about any other number. This means that the highest point of the curve shown in Figure II corresponds to the average of the table from which the curve was constructed. In a skew distribution the average value and the highest point of the curve, the mode, do not coincide, as indicated in Figure III.

Perhaps we can best illustrate the important features of a statistical table, or frequency distribution, by using an example from the realm of games of chance, approached this time from the statistical or empirical point of view. Table XXIV gives the results of a widely known experiment in dice throwing made by W. F. R. Weldon, which was conducted as follows: Twelve dice were thrown 4,096 times; when the 4, 5, or 6 came up on any die, it was entered as one success, and the number of successes in each throw was recorded.

TABLE XXIV

Number of successes:	0	1	2	3	4	5	6	7	8	9	10	11	12	Total
Number of throws:	0	7	60	198	430	731	948	847	536	257	71	11	0	4,096
Theoretical number of throws:	1	12	66	220	495	792	924	792	495	220	66	12	1	4,096

For example, the fourth column of the table reads three successes against 198 throws; this means that on 198 of the 4,096 throws, exactly three of the twelve dice showed a 4, 5, or 6.

The average number of successes, as computed from the table, is 6.139. We shall pass over, for the time being, the details of this calculation. The method of computing the average of a frequency distribution, in general, will be illustrated in the next chapter. But we can also compute this average without reference to experiment, since we are dealing with the simple game of dice. It is ½ × 12, or 6.000; the discrepancy between theory and practice is 0.139.

The proportion of successes, in other words the probability of success, indicated by the experiment is 6.139/12, or 0.512. The theoretical chance of success is evidently ½, or 0.500 (three favorable cases out of a total of six cases). The discrepancy between theory and experiment is 0.012.

This comparison between theory and practice, as far as it goes, seems to indicate at least a fairly close accord. But we are far from having a real measure of the agreement between the two sets of figures, and we are not yet prepared to answer the question, Are the results of the Weldon dice-throwing experiment in accord with the laws of chance? The answers to questions of this sort are of the greatest importance in many statistical problems, as we have already seen, and it is time that we indicate in more detail one of the methods that the statistician uses in attacking such problems.

We have studied the Weldon experiment up to this point as a frequency distribution consisting of 4,096 repetitions of a certain event, namely throwing twelve dice at a time. We have computed the average of the table, which tells us the average number of successes per throw, and from it the statistical probability of success, as indicated by the experiment. Both these quantities, as we have seen, are slightly in excess of the theoretical figures. But we have observed several times, in these pages, that the throw of each die is independent of that of the others, and therefore that throwing twelve dice once is the same thing, as far as the theory of chance is concerned, as throwing one die twelve times. It follows that we can equally well regard the Weldon experiment as consisting of 12 × 4,096, or 49,152 throws of a single die. As the chance of success on each throw is ½, the most probable number of successes is ½ × 49,-152, or 24,576. Let us first of all compare this with the actual number of successes observed. To find the latter we must multiply each number in the first line of Table XXIV by the corresponding number from the second line. This gives $(0 \times 0) +$ $(1 \times 7) + (2 \times 60) + \ldots$ and so on, and the total is 25,145, which is a deviation from the expected value of 569 in excess.

Our problem is this: Is a deviation of 569, or 2.3 per cent, one that might be expected to occur, according to the laws of chance, reasonably often, or is it so rare a phenomenon as to constitute one of those "unbelievable coincidences" referred to in Chapter IV?

In order to find the answer we must first accurately define and compute the *standard deviation* which, we recall, measures the scattering or dispersion of a set of numbers about its average. As the standard deviation is the square root of a certain average, we define first its square. *The square of the standard deviation is the average of the squares of the deviations of the numbers in the set from their arithmetical average, or mean.* The positive square root of the quantity so computed is the standard deviation itself.

A simple example will make this clear. Suppose that there are four men in a room whose heights are 65, 67, 68, and 72 inches. The average of their heights is 68 inches. The deviations of the heights from this average are respectively —3, —1, 0, and +4. The squares of the deviations are +9, +1, 0, and +16, and their sum is 26. Their average is one fourth of 26, or 6.5. The standard deviation is therefore the positive square root of 6.5, which is equal to 2.55.

Now that we know how to compute the value of the standard deviation, we are close to the answer to our question. Our next step requires us to make a new assumption, however, regarding our experimental data. We must assume that the frequency distribution is one that closely follows the normal distribution (Figure II on page 226). If this is not true, the method that we are following requires modification. But if the data represent a series of independent events with constant probability, as in throwing dice or tossing coins, it is proved mathematically that the distribution is close to normal, provided that the number of trials is large, and that the probability of success in each is not too small. We are therefore safe in proceeding with our study of the Weldon experiment.

Under these conditions a quantity known as the *probable error* is of great statistical significance, and it is found from the standard deviation by multiplying the latter by $\frac{2}{3}$ (more accurately 0.6745). Its significance and value are due to the following fact: The probable error is a deviation in either direction from the average such that the betting odds that a deviation, selected at random, will exceed it are even. If, then, we find in an experiment like that of Weldon that the total number of successes differs from the expected number by two thirds of the standard deviation, as computed from the ideal frequency table furnished by the theory of the experiments, we shall know that in the long run such a deviation should happen about one time in two. But the odds against a deviation greater than a given multiple of the probable error increase very rapidly with increasing values of the multiple, as shown in the following table:*

TABLE XXV

Multiple of Probable Error (plus or minus)	Odds against a Larger Deviation from the Expected Value
1	1 –1
2	4.5–1
3	21 –1
4	142 –1
5	1,310 –1
6	19,200 –1
7	420,000 –1
8	17,000,000 –1
9	100,000,000 –1

* H. C. Carver, *Handbook of Mathematical Statistics* (H. L. Rietz, Ed.), Houghton Mifflin Company, Boston, 1924, p. 100.

The first line of this table merely repeats the definition of the probable error as that deviation (in either direction) on which it would be fair to bet even money. The second line, however, tells us something new. It states that the probability that a random deviation will exceed (in either direction) *twice* the probable error is 1/5.5 (odds 4.5—1 against). And the last line states that there is only 1 chance in about 100,000,000 that the deviation will exceed *nine* times the probable error.

This table is in reality an extract from a more complete one, that gives the probabilities of deviations of fractional multiples of the probable error as well as whole-number multiples of it. But it is sufficient to show us how very rapidly the probability of a deviation decreases with the size of the deviation.

We need now to know how large the probable error is, as predicted by the theory of chance, for a series of 49,152 throws of a die, the probability of success on each throw being $\frac{1}{2}$.

Fortunately, this has been worked out for us by the mathematicians, and we need only take their results and apply them to the case before us. Here is what they have found: If you make a series of any number of trials, say n, each with probability p of success, q of failure ($p + q = 1$), and if these trials are independent, as they are in rolling dice or tossing coins, the probabilities of various numbers of successes and failures are given by the terms of $(q + p)^n$. To find these terms you multiply $(q + p)$ by itself n times, according to the rules of algebra. The first term is q^n, and it tells you the probability of 0 successes (n failures). The second term is npq^{n-1}, and it tells you the probability of getting exactly 1 success and $n-1$ failures, and so on for the other terms. Let us illustrate just how this works for a simple example. Suppose that we toss a coin four times, and ask the probabilities of various numbers of heads, so that a success means heads. As we have just learned, these probabilities are given by the terms of $(q + p)^4$. Here both q and p are equal to $\frac{1}{2}$, so that we have

$$(\frac{1}{2} + \frac{1}{2})^4 = \frac{1}{16} + \frac{1}{4} + \frac{3}{8} + \frac{1}{4} + \frac{1}{16}.$$

The probabilities we wish are given respectively by the terms on the right side of this equation, as shown in the following table:

No. of Heads	Probability
0	$\frac{1}{16}$
1	$\frac{1}{4}$
2	$\frac{3}{8}$
3	$\frac{1}{4}$
4	$\frac{1}{16}$

Notice that the sum of the probabilities is 1, as it must be. You will find it easy to verify these results according to the principles of Chapters V and VI.

Now if you repeat this series of n trials a number of times, say N, the frequencies of occurrence of the various numbers of successes are given by the terms of $N (q + p)^n$. This gives us, then, the theoretical frequency distribution that applies to any experiment of the nature of the Weldon experiment, which we are studying. Notice that since $p + q = 1$, the sum of all the terms is exactly N, as it must be.

We wish to know the average and the standard deviation of the frequency distribution given by $N (q + p)^n$. Since it is easily shown that the value of N makes no difference, we can as well think of it as 1 and ask the same questions for $(q + p)^n$. The solution of the problem is easily found by mathematical methods, keeping in mind the definitions of the average and the standard deviation, and we shall merely state the results. The average of the distribution, which coincides with what we have called the expected number of successes, is np, and the formula for the standard deviation is $\sqrt{npq}$. Since symbols, not numbers, have been used, these results are general and can be applied to any case, regardless of the values of p, q, and n. We shall appreciate this fact when we apply our findings to the

Weldon experiment, for the frequency distribution is then given by the terms of $(\frac{1}{2} + \frac{1}{2})^{49,152}$, and the direct numerical calculation of the average and the standard deviation would therefore involve 49,153 terms. This is a small example of the power of mathematical methods.

We may note, incidentally, that the theoretical distribution given in the third line of Table XXIV, which applies to the experiment in the form there recorded, is obtained from the terms of 4,096 $(\frac{1}{2} + \frac{1}{2})^{12}$. The exponent 12 is the number of dice thrown on each of the 4,096 trials.

Now that we have an expression for the value of the standard deviation, we at once obtain that for the probable error. It is 0.6745 $\sqrt{npq}$.

We are finally in a position to return to the Weldon dice experiment, and to compute the probable error that the theory of chance specifies. According to the above formula it is 0.6745 $\sqrt{49,152 \times \frac{1}{2} \times \frac{1}{2}}$, which is equal to 74.8. The chance that a random deviation from the expected number of successes, 24,576, will be more than 75 is therefore about even. What is the chance that an experiment will result in a deviation of 569, as did that of Weldon? This is a deviation of 7.6 times the probable error, and a glance at Table XXV tells us that the odds against such an occurrence are roughly several million to 1. The actual odds are about 4,500,000 to 1.

What are we to conclude from this extraordinary result? That the laws of chance are incorrect? Certainly not, for in this experiment we have been dealing with mechanical contrivances, namely a set of dice, and in order to be able to test the laws of chance we should have to be assured in advance that the dice were perfectly true, or, at least, they had a perfectly uniform bias, so that the chance of success would be constant. Now the only practical way to test the accuracy of dice, which are always under suspicion due to the varying number of spots on different sides, is to compare the results of rolling them with the results predicted by the laws of chance. We are back precisely to where we started, and we see clearly that experiments in rolling dice can test only the trueness of the dice. We shall have to turn in another direction if we wish to test the laws of chance.

We are still faced with the fact, however, that Table XXIV, looked at as a frequency distribution, yields an average (and, as a matter of fact, a standard deviation) that is in fairly good accord with the pure chance values. Our conclusion must be that the dice used in the experiment tended to turn up a slightly disproportionate number of fours, fives, and sixes, as a group, but that the *distribution* of successes by trials was in close accord with the laws of chance. A glance at the third line of the table, which gives the theoretical frequency distribution, computed according to the laws of chance, tends to bear out this view. You will notice an excess of throws in which the number of successes is from 6 to 10, inclusive, and a corresponding deficit where the number of successes is from 0 to 5, inclusive. The effect of this is to displace the actual frequency curve, as compared with the theoretical, somewhat to the right (the direction of increasing number of successes). In other words, the distribution of successes is of the type anticipated, but the probability of success on each trial is not exactly $\frac{1}{2}$.*

---

* Another of the Weldon dice-throwing experiments, very similar to that studied in the text, is discussed by Sir Ronald A. Fisher in his *Statistical Methods for Research Workers,* 6th edition, page 67, one of the leading technical works on the subject. In this instance twelve dice were thrown 26,306 times, and on each throw the number of 5's and 6's showing were listed as successes, thus leading to a frequency table altogether similar to our Table XXIV. The conclusion reached by Sir Ronald is that the results obtained would happen on the pure chance hypothesis (the dice being assumed true) only once in five million times. This conclusion is so close to that reached in the text that one can suspect that the same set of dice were used in the two experiments. Or possibly all dice have a very small bias, due to the different numbers of spots, which shows up only in very long series of trials.

Since mechanical contrivances, coins, dice, roulette wheels, and so on, are subject to mechanical biases, we shall turn, for our next illustration of these statistical procedures, to an example of another sort. In drawing cards from a pack the problem of avoiding possible biases is very much simplified, so that we can feel confident that the results obtained by a skillful experimenter should agree with those predicted by the laws of chance. One of the leading modern statisticians, C. V. L. Charlier, made 10,000 drawings from an ordinary deck, recording as a success each drawing of a black card. The total number of black cards drawn was 4,933, giving a deviation of 67 from the expected number, which is of course 5,000. Applying our formula for the probable error we obtain $0.6745 \sqrt{10,000 \times \frac{1}{2} \times \frac{1}{2}}$, which equals 0.6745×50, or 33.7. The actual deviation is, then, almost exactly twice the probable error, and Table XXV tells us that we should expect such a deviation about one time in 5.5, in the long run. The result is therefore entirely in accord with that predicted by chance.

This experiment was performed by Charlier with a slightly different object in view. The drawings were recorded in sample sets of ten, count being kept of the number of successes in each.* In this way a frequency table was obtained, similar to that in Table XXIV. Both these tables represent the type of frequency distribution that we have just studied and are known as Bernoulli series. They are characterized by the repetition of independent chance events, the probability of success remaining the same in each. They are so named because the mathematician James Bernoulli made some important discoveries regarding them.

There are many other quantities, in addition to the average and the standard deviation, that can be computed from a

* The experiment is reported in full in Arne Fisher: *The Mathematical Theory of Probabilities*, 2nd ed., 1930, p. 138.

table. Each one helps to characterize the table as a whole; each one is shorthand for some important property of the distribution. But it is to be kept in mind that no two, or three, or four characteristics of a statistical table tell its whole story, unless the distribution is one that is known to follow a simple statistical law, a relatively rare occurrence. Even in that event, it is not the table itself that is represented, but the ideal form that the table would assume, if its data were infinitely numerous. This ideal form is in itself of great value, and to find it is one of the objects of many statistical studies. It is of no value, however, until it has been definitely ascertained that the discrepancies between it and the experience table can reasonably be due to chance fluctuations.

The theory of these theoretical frequency distributions is at the root of an important section of mathematical statistics. For the results apply not only to rolling dice, but to any series of trials that are independent in the probability sense of the word, and in which there is a constant probability of success from trial to trial. It is thus possible, by comparing these Bernoullian frequency distributions with others based on empirical statistics, to test the hypothesis that a constant probability lies behind the latter.

In addition to the Bernoullian distribution, or curve, there are two others that are closely related to simple games of chance, and that are of value to the statistician in analyzing statistical material of various sorts. One is known as the Poisson distribution, the other as the Lexis distribution. These three distributions are usually thought of as arising from drawing white and black balls from several urns. We get a Bernoulli distribution if each urn has the same proportion of white and black balls, so that the probability of drawing a white (or black) ball is the same in each urn, and if each ball is replaced before the next drawing from that urn takes place (it comes to the same thing to use a single urn). We get a Poisson distribution if the proportion of white and black balls varies from urn to urn, so that the probability of each varies, and if each ball drawn is replaced at once. We get a Lexis distribution if we make each set of drawings exactly as in the Bernoulli case, but change the proportions of the white and black balls between sets of drawings.

Each of these three cases represents a probability situation that can be solved, like those we met with in poker and roulette. On the other hand, the sets of drawings may be thought of as statistical series or frequency tables. In each case the theoretical average and standard deviation can be computed, and they serve as excellent criteria for judging and classifying the statistical series met with in practice. It is clear that in this way the ideas of the theory of chance are brought into the most intimate relation to statistics.

CHAPTER XVI

Fallacies in Statistics

To WRITERS of popular articles and books on serious or semi-serious subjects, in which each contention is vigorously proved, statistics are an unadulterated boon. As their author is not ordinarily a professional statistician, he finds his figures pliant and adaptable, always ready to help him over the rough spots in an argument and to lead him triumphantly to the desired conclusion. He handles the statistical data with ease, takes a couple of averages of this or that, and the proper conclusions fairly jump out of the page. He may never have heard of dispersion, or correlation coefficients, or of a long list of other things; nor has he any clear idea of the necessary criteria for selecting the material in the first place. These omissions are bound to simplify statistical studies enormously, and the practice would be followed by all statisticians if it were not unfortunately the fact that it leads almost inevitably to errors of all descriptions.

Indulging in statistical analyses without technical preparation is much like flying an airplane without training as a pilot. If you can once get safely off the ground, you will probably be able to fly along on an even keel quite satisfactorily, provided that nothing out of the way happens. It is when you decide to end the flight and start down that your deficiencies become evident. Similarly in statistical investigations it is the start and the finish that present difficulties, that is, collecting or selecting the data and drawing the conclusions. What remains can be done by almost anyone, once he understands exactly what is to be done.

It is not essential to be a technically trained statistician in order to distinguish good statistics from bad; for bad statistics has a way of being very bad indeed, while a good analysis carries with it certain earmarks that can be recognized without great difficulty. In a good job the sources of the data are given, possible causes of error are discussed and deficiencies pointed out; conclusions are stated with caution and limited in scope; if they are numerical conclusions, they are stated in terms of chance, which means that the probabilities of errors of various amounts are indicated. In a bad job, there is a minimum of such supplementary discussion and explanation. Conclusions are usually stated as though they were definite and final. In fact, it can be said that a poor job is characterized by an air of ease and simplicity, likely to beguile the unwary reader into a false sense of confidence in its soundness. In this respect statistics is very different from other branches of mathematics, where ease and simplicity are earmarks of soundness and mastery.

We have already stressed the extreme importance of studying the source of the data that enter into a statistical table, or of selecting the material, if that is possible, with a view to the use to which it is to be put. The group of individuals (of any kind) must be a fair sample of the whole "population" that is being studied. What this means, precisely, is usually clear enough in any particular case, but it is very difficult to formulate anything like a set of rules to apply to every case.

Suppose that we are interested in collecting statistics bearing on typhoid fever, with a view to formulating conclusions applying to the country as a whole. We shall find it extremely difficult to decide what constitutes a fair sample, for the reason that the incidence of typhoid in a region depends largely on the sanitary conditions prevailing in that region. In order to select a set of areas suitable to represent, in this question, the country as a whole, it is necessary first to have an accurate gauge of the sanitary conditions in these sample areas and of their relation to those of the country at large. This requires an entirely independent investigation which must precede the main one.

As a matter of fact, the competent statistician would very soon discover from the table of typhoid statistics from various regions, even if he did not know that the figures concerned a contagious disease, that there was some disturbing element present, so that his results would not be valid without further investigations.

The method that the statistician uses to attain such results is based on the comparison of the actual frequency distribution with ideal or theoretical distributions like those of Bernoulli, Poisson, and Lexis (and some others) which were mentioned at the end of Chapter XV. These ideal distributions are simply those that arise in various games of chance, where the probability of an event remains the same from one trial to another or changes from trial to trial in some comparatively simple manner. This demonstrates again how closely the theory of statistics is united to the theory of chance and how impossible it would be to understand statistical methods or interpret statistical results without first understanding something of the theory of chance.

The most used, and therefore the most misused, of statistical ideas is the ordinary arithmetical average. Almost everyone knows how to take a simple average of a set of numbers. The average is the sum of the numbers divided by the number of numbers in the set. If the average is desired of a set of measurements of distances, for instance, the distances may be in opposite directions, in which case those in one direction are marked *plus,* those in the other *minus.* In finding the average of such a set, the arithmetical sum of the negative numbers is to be subtracted from the sum of the positive numbers, giving the sum of all the numbers. The latter is then divided by the number of numbers, giving the average.

The method of taking the average of a frequency distribution is best illustrated by an example. In an examination given to thirty-five students the papers were marked on a scale of 10, and the number of students obtaining each of the grades was recorded as follows:

TABLE XXVI

Grade	Number of Students (Frequency)	Product
10	1	10
9	3	27
8	12	96
7	10	70
6	6	36
5	2	10
4	0	0
3	1	3
Total	35	252

Thus one student got a grade of 10, or 100 per cent, three got a grade of 9, or 90 per cent, and so on. The third column, headed *Product,* was obtained by multiplying the corresponding numbers in the first and second columns. The average grade of the thirty-five students is found by dividing the total of the third column, which is 252, by the total of the second column, which is 35, giving 252/35, or 7.2, that is to say 72 per cent. This is the method that was used, for example, in computing the average from Table XXIV, in the preceding chapter.

This averaging process can be looked at in this way: The thirty-five students made a total point score of 252 on the examination, so that the score per student was 7.2 points. This seems clear and simple enough. Yet it is true that many people make serious blunders in taking the average, when the frequency of the numbers to be averaged must be taken into account. Very often the average is computed as though all the frequencies were the same. For the example of the examination grades this would correspond to computing the average by adding together the grades, which gives 48 (notice that the grade 4 is omitted, since no student received it), and dividing by their number, which is 7. This gives for the average $48/7$, or 6.86, and the error committed is —0.34, about one third of a point.

Or take a simple illustration from advertising statistics. Suppose that one newspaper, call it A, sells its advertising space for exactly four times as much as newspaper B charges, and that A's daily circulation is twice that of B. If an advertiser uses the same amount of space in each (it does not matter what the amount is), what is the average rate that he pays? Here the frequencies are the two circulations, and we see that the average rate is exactly three times the rate charged by B (or three fourths of that charged by A). This is the average of the rates themselves. But in advertising practice it is convenient to use rates defined in such a way as to be independent of the page size and circulation of the newspaper or magazine in question. This is accomplished by using the rate per line of advertising per unit of circulation. The unit of circulation usually adopted is 1,000,000. To average two or more such line rates, we follow the same procedure; each is multiplied by the corresponding circulation, and the sum of the products is divided by the sum of the circulations. In our example of newspapers A and B, the average of the line rates turns out to be $5/3$ the line rate of B, or $5/6$ that of A. If we made the error of leaving out of account the frequencies (circulations), the average would be $3/2$ the line rate of B.

If many blunders of this sort are committed, it is also true that many of those who commit them are thoroughly at home in dealing with averages that have to do with one of their hobbies or interests. Take the baseball averages, for instance. Millions of baseball fans follow these averages from day to day, and know how to compute them themselves, if the occasion arises. They would not think of trying to find a player's batting average, for example, by adding together his batting averages for each day of the season and dividing the result by the number of games. Instead they would divide his total number of hits for the season (or any given period) by the number of times he was at bat.

Apart from errors in computing the average of a set of numbers, one encounters many mistaken notions concerning the significance and use of the idea of averages. This is true even among technical writers in various fields. In his interesting book *The Human Mind,* published many years ago, Dr. Karl A. Menninger says (page 190) : "Fortunately for the world, and thanks to the statisticians, there are as many persons whose intelligence is above the average as there are persons whose intelligence is below the average. Calamity howlers, ignorant of arithmetic, are heard from time to time proclaiming that two-thirds of the people have less than average intelligence, failing to recognize the self-contradiction of the statement."

This naïve belief that half the members of a group are necessarily above, and the other half necessarily below the average of the group is refuted by the simplest possible example: There are four boys in a room, three of whom weigh 60 pounds, the fourth 100 pounds. The average weight of the four boys is 70 pounds; three of the four are below the average, only one above. The "calamity howlers" may be entirely mistaken in their assertion that two thirds of the people have less than average intelligence, but they must be silenced by statistical counts, not by erroneous reasoning.

The error just referred to amounts to assuming that the average of a frequency distribution always coincides with what is called the *median* of the distribution. The median can be thought of as a line drawn through a frequency table dividing it into two parts, such that the sum of the frequencies in one part is equal to the sum of the frequencies in the other.

In terms of a frequency curve, such as that shown in Figure III, page 230, the median is the vertical line that divides the area under the curve into two equal parts. If the frequency distribution is one that follows Gauss's law, Figure II, page 226, or is symmetrical about the average, then the median and the average coincide. Barring these cases, which are relatively rare in statistical practice, they do not coincide. When the distribution is not far from normal, however, the average and the median are close together. In the above example of the examination grades, for instance, the distribution is only mildly skew; the median comes out to be 7.35, thus differing from the average by only 0.15.

There is a second and somewhat similar error concerning the average of a distribution that is sometimes committed and which may also have its origin in thinking of distributions in general as though they all followed Gauss's law. This consists in assuming that more frequencies in the distribution correspond to the average value than to any other value. This error somewhat resembles the one in the theory of chance, mentioned in Chapter X, which consists in believing that the most probable single event of the possible events is more likely to occur than not. In the example of the students' grades, the greatest frequency corresponds to the grade 8, while the average, as we have seen, is 7.2. The difference amounts to 0.8.

These confusions regarding the arithmetical average are easily avoided by keeping in mind the fact that it is but one of the characteristics of a distribution, and that the distribution can be changed in countless ways, without changing the value of the average. For example, we can replace the distribution of students' grades just cited by the following one, and it will be seen that the average remains at 7.2:

TABLE XXVII

Grade	Number of Students (Frequency)	Product
10	1	10
9	7	63
8	8	64
7	10	70
6	3	18
5	3	15
4	3	12
Total	35	252

Average = 252/35 = 7.2.

Although this distribution has the same *average* as the preceding one, the largest single frequency in it is 10, and corresponds to the grade 7, while in the other we see that the largest frequency is 12, corresponding to the grade 8. When we represent a frequency distribution by a smooth curve, like those in Figures II and III, the vertical line through the highest point of the curve indicates the value to which corresponds the greatest frequency; this value is called the *mode*.

We have tacitly assumed that a frequency distribution starts with a low frequency, increases to a maximum, and then falls to a low value again. This is true of a large number of distributions, including all those that we have mentioned so far, but it is by no means true in general. A frequency curve may have several maximum values; each such value is called a *mode*. This fact alone indicates the fallacy of assuming that the mode and the average coincide. A frequency distribution, no matter how irregular in appearance, can have but one arithmetical average.

Of great importance is what is known as the *weighted average*. Each of the numbers to be averaged is given a weight, which is also a number. If all the weights were equal, the weighted and the unweighted average would be the same. The average of a frequency distribution, such as those we have taken above, is in reality an example of a weighted average; the frequencies are the weights. In the above example of the students' grades, for instance, the weights are the number of students corresponding to a certain grade, and the average of the frequency distribution is the same as the average of the grades, each being weighted with the number of students who obtained that grade.

It is clear that the greater the weight that is attached to a given number in a table, the greater the influence of that number on the average of the table and vice versa. Weighted averages are of great value in the theory of measurements. In many cases the measurements are not equally reliable, due, for example, to variations in the conditions under which they were made. If each measurement is weighted in proportion to its estimated quality, the resulting weighted average gives an improved estimate of the quantity measured. It must be kept in mind that there is no law compelling us to adopt one kind of average rather than another. Our sole object is to deduce from the set of measurements as good an estimate as possible of what we are attempting to measure. If all the measures were in accord with one another, there would be no question as to which value to adopt. When they are not in accord, it is the object of statistical theory to indicate the method of choice and the reliability of the value adopted.

The weighted average is used not only for tables of measurements, but for data of any kind where some parts of the material are thought to be more reliable than other parts, or where, for some legitimate reason, some parts are to be given particular emphasis. Clearly, the value of the weighted average in a particular problem depends altogether upon the weights to be used, and this choice is capable of making or breaking the statistical investigation in which it occurs.

What has been said so far about the use of the weighted average is based on the assumption of intellectually honest statistics. For the other kind of statistics weighting is indeed a powerful weapon. An average can be pushed up or pushed down, a conclusion can be toned up or toned down, by the simple expedient of assigning judiciously selected weights to the various parts of the data. When you encounter statistical analyses in which data have been weighted, try to determine, before swallowing the conclusions, whether or not the weighting has been done according to some objective rule that was fixed in advance, preferably before the investigation was undertaken. Failing in this, you would do well to preserve a skeptical attitude.

As an illustration of a case where a weighted average is badly needed we draw attention to the usual listing of best-selling books. These lists are intended as an index to the current reading taste of the American public and are based on reports from a large number of bookstores in various parts of the country. The rating of a book is determined by the number of bookstores that report it on their list of best sellers, and no account whatever is taken of the relative number of sales at the different stores. Thus the smallest bookstore is given the same weight as the largest. The lack of correct weighting makes the reports a biased sample of the reading tastes of the country.

In spite of the simplicity and familiarity of the ideas of averages, fallacious results are sometimes reached by averaging the wrong things. For example, ask a few people the following question and note what percentage of the answers are correct: If you drive your car ten miles at forty miles per hour, and a second ten miles at sixty miles per hour, what is your average speed for the entire distance? Many of the answers will be fifty miles per hour, which is the average of forty and sixty. But this is not correct. You have driven your car a total of twenty miles, the first ten miles requiring fifteen minutes, the second ten miles requiring ten minutes. You have driven twenty miles in a total of twenty-five minutes, which means that you have averaged forty-eight miles per hour, not fifty miles per hour. Furthermore, the correct average is always less than the average of the two speeds, a result easily proved mathematically.*

In this calculation it is essential that the two distances be equal, but it does not matter what this distance is, whether one mile, ten miles, or one hundred miles.

After finishing this experiment try asking the same people the following: If you drive your car for one hour at forty miles per hour and for a second hour at sixty miles per hour, what is your average speed for the entire time? In this instance it is correct to average the speeds, and the answer to the question is fifty miles per hour. If the two times are *not* equal, the correct average is found by taking the *weighted average,* the speeds playing the role of weights.†

In discussing the probabilities of various events in a series of events, such as tossing coins or rolling dice, we can state our results in two forms. We can say that the probability of throwing the double six with two dice is $\frac{1}{36}$, or we can say that in a long series of throws of two dice the double six will come up one time in thirty-six, *on the average.* It is sometimes rather thoughtlessly concluded from the latter form of the

* Here is the answer to this problem in mathematical form. If you drive d_1 miles at the speed v_1 and d_2 miles at speed v_2, your average speed, V, for the entire distance is $V = \dfrac{v_1 v_2 (d_1 + d_2)}{d_1 v_2 + d_2 v_1}$. If the two distances are equal the average speed is $V = \dfrac{2v_1 v_2}{v_1 + v_2}$, and it is very easy to prove that V is always less than $\frac{1}{2} (v_1 + v_2)$, unless the two speeds are equal, in which case V is equal to their common value.

† In this case, if we denote the two times by t_1 and t_2, the average speed, V, is easily seen to be $V = \dfrac{v_1 t_1 + v_2 t_2}{t_1 + t_2}$. This is the weighted average referred to in the text. If the two times are equal to each other, we have $V = \frac{1}{2} (v_1 + v_2)$. This is the simple arithmetical average of the speeds, as stated.

statement that if an event has been shown to occur on an average of once in n trials, it is certain to occur at least once in every n trials. This is far from the case, as we saw in Chapter XII in connection with roulette. But as the series of trials becomes longer and longer, it is of course true that the probability that the event will take place at least once approaches nearer and nearer to unity, which represents certainty. In thirty-six rolls of two dice, for example, the probability that the double six will appear is 0.637, or a little less than two chances out of three. When we are dealing with a general statistical situation we usually have to do with empirical or statistical probabilities, instead of a priori probabilities, but the reasoning in all these matters remains the same.

The type of error that we have just called attention to reminds us of the story of the surgeon who was about to perform a very dangerous operation on a patient. Before the operation the patient came to him and asked to be told frankly what his chances were. "Why," replied the doctor, "your chances are perfect. It has been proved statistically that in this operation the mortality is 99 out of 100. But since, fortunately for you, my last ninety-nine cases were lost, and you are the one hundredth, you need have no concern whatever."

We have discussed two kinds of average, the simple arithmetical average and the weighted arithmetical average. There are many other kinds of average that are used in statistics, some of which are of great importance. An average is merely an arbitrary way of combining a number of numbers so as to obtain a single number that for some particular purpose can be used to represent the entire set, or to summarize certain of its properties; so there is no limit to the number of possible kinds of average of a set of numbers. One could go on forever inventing new ones, but it would be a futile occupation unless some use could be found for them.

The *geometric average,* or *geometric mean,* of two numbers is defined as the positive square root of their product. For example, the geometric average of 2 and 8 is the positive square root of 16, or 4. The arithmetic average of 2 and 8 is 5. The geometric average of three numbers is the cube root of their product, and so on. One of the practical uses for the geometric average is found in estimating the size of a population midway between two dates on which it was known, on the assumption that the *percentage* increase in the population is the same each year. If it were assumed that the absolute increase in the population is the same each year, then the ordinary arithmetical average would give the better estimate. The geometric average is used likewise in estimates concerning all quantities that increase according to the law of compound interest. An example from economics is total basic production over certain periods of years. Hereafter, when the word *average* is used without qualification, we shall understand, following common practice, that the ordinary arithmetical average is indicated.

Since many of the important quantities used in statistics, such as the standard deviation, are in reality averages of something or other, it is correct to say that the *idea* of averaging is fundamental in statistics. But this does not justify us in concentrating all our attention on the average of a frequency distribution, to the exclusion of these other important quantities, and of the distribution as a whole. Except when the data are of the crudest sort, it is also necessary to have a measure of the reliability of the average, as computed, and this again involves a consideration of the dispersion. An example of a case where the average is an important thing is any set of measurements that are subject to accidental errors. The most probable value is then the average. The *probable error,* which we have defined in the last chapter as that deviation of an individual value from the average, on which it would be fair to bet even money, is a measure of the reliability of this average. It is essential to attach the probable error to each set of measurements, and no reputable precision measurements are published without it.

In other cases the practical value of the average depends greatly on our particular point of view. The quartermaster of an army may report, for example, that the average food consumption per soldier during a certain campaign was ample. If it is subsequently discovered, however, that some of the men were starving, while others received far more than the average, the report becomes not only inadequate but misleading. From the point of view of the individual soldier, the vital thing is an equitable distribution of food, not the average consumption. And this requires that the dispersion be small.

Another simple illustration comes from the well-known fact that the gasoline consumption of a motor depends upon the speed at which it is driven. The amount consumed in an automobile per mile decreases with increasing speeds until a minimum is reached; further increases of speed result in a higher rate of consumption. When you are touring, therefore, your average speed does not determine the rate at which you are using gasoline. You can maintain an average speed of forty miles per hour in many different ways. You may drive very fast part of the time and very slowly in between, or you may keep the variations in speed within narrow limits. To check your gasoline mileage accurately, what you need is a frequency distribution of speeds, that is to say, a table giving the length of time you drove at each speed.

On the other hand, for most purposes it is altogether sufficient to know your average speed. As far as the hour of reaching your destination is concerned, it makes no difference what the distribution of speeds is, provided that the average is the same. If you wish to know, however, your chance of gaining a half hour (or of losing it), you must turn to your previous records of touring and compute from them the standard deviation, which measures the probability of fluctuations about the average.

When the values of a frequency distribution are closely packed about the average value (in other words, when the dispersion is small) this average value may be considered as fairly representing the class of individuals described by the frequency distribution. When, however, these values are widely scattered about the average, or when they are closely packed about several widely separated values (in other words, when the dispersion is large), the average value does not fairly represent the class in question, and its unqualified use may be extremely misleading.

In certain cases this failure of the average to be of significance may be due to the fact that several distinct classes of individuals have been scrambled together in the frequency distribution. We should get such a distribution, for example, if we put into one table heights of Americans of both sexes. The average height indicated by such a table would correspond to neither men nor women but would fall between the two "cluster points" that correspond to the average heights of men and of women. The distribution, furthermore, would not follow the law of Gauss, as does that for either of the sexes separately. Similarly, if we put into one table heights of men of various races, we should obtain a still more complicated frequency distribution, and the average would have still less significance. Evidently it is of great importance to unscramble these mixtures of distinct classes of individuals, whenever it is possible to do so. Sometimes this can be reliably accomplished through a study of the frequency distribution itself, rather than through an analysis of the sources of the data; but this is the exception, not the rule. If we have to do with the sales of a business, there may be many distinct classes of customers involved and it may be of great value to sort them out. We shall come across cases of the sort when we turn our attention in a later chapter to the role of chance and statistics in business.

There are cases where the values in a frequency table are so scattered that the ordinary average is simply meaningless. And yet some individuals have so strong an urge to take the average that even in such a case they cannot resist the temptation to do so. They seem to be beguiled by the fact that the average exists. Now the average is merely the sum of the numbers divided by the number of numbers, and to say that the average exists is equivalent to saying that the sum exists, which it always does if the number of frequencies is finite. That is a different matter from saying that the average is significant. A group of telephone numbers always has an average, but it is hard to know what one would do with it. A case of this sort occurs in the chapter on blood pressure in one of the leading medical books for home use. The usual charts are given, showing average systolic and diastolic pressures by ages, and arithmetic formulas are included. They show the average systolic pressure ranging from 120 at age twenty to 138 at age seventy. On the following page, however, we find that a random check of several hundred office patients disclosed that less than one third had systolic pressures between 110 and 140, roughly the same number over 140, and the balance below 110. Unless this was an extraordinarily biased sample it indicates that the probability of having a reading within ten or twenty points of the "average," regardless of age, is quite small. Under such circumstances why use the word *average* at all?

It is possible that this situation could be repaired, like some of those referred to in a preceding paragraph, by demonstrating that individuals fall, as regards blood pressure, into several well-defined groups, each showing correlation between pressure and age, and each showing small dispersion about the averages. These conditions are essential. Otherwise the reclassification would be meaningless.

There is a type of statistical fallacy that is less obvious than those concerning the average, but for that very reason more insidious. It has its roots in an incomplete or faulty analysis of the factors that enter a statistical situation. For an example we turn again to medicine.

The statement has frequently been made in the press and on the platform that the deathrate from cancer is increasing, and it is quite correct that the number of deaths attributed to cancer per thousand of population has increased materially in recent years. The real problem, however, is not what the statistical table shows, but what it means. Is it true, as is so frequently implied, that the frequency of death from cancer is actually increasing?

In order to be able to make such a statement with accuracy, the following factors, and possibly some others, must be taken into account: First of all, cancer is now more accurately diagnosed than before, with the result that many deaths formerly listed under other diseases or under "cause unknown" are now listed as due to cancer. Second, there is an increased number of post-mortem diagnoses, due to the increased number of autopsies. Third, great improvements have been effected in the accurate reporting of cases of deaths and in collecting such data. Fourth, the average length of life has increased, due above all to the decrease in infant mortality. As cancer is primarily a disease of later life, there has been a correspondingly increased proportion of the population "exposed" to it. When all these factors have been properly measured, and their effects taken account of, it may indeed turn out that the true death rate from cancer is increasing. But such a conclusion, based merely on the fact that the number of deaths per thousand of population attributed to cancer has increased, is bad statistics.

One of the much discussed questions of the day is the relative safety of passenger travel by air. If we seek to inform ourselves on this subject, we are met with a barrage of statistics. Since airplanes fly at high speed and therefore cover a large number of miles, those who wish to minimize the risks of air travel can give the number of deaths or injuries per mile per passenger (or per occupant), that is to say, per passenger mile. Those who wish to maximize the risk in air travel can base their case on the number of deaths or accidents per passenger hour. What is the correct way to measure the rate of air fatalities? Or we might ask, Is there such a thing as a correct way?

Suppose that you are going to make a nonstop flight from New York to Washington, which is a distance of roughly 180 air miles. You are perhaps interested in determining the risk that you take (although it is better to indulge in such thoughts after the flight is over). To determine your risk, we need the best available records of previous flights between New York and Washington. From these records we must discard, on a common-sense basis, those flights which, for one reason or another (such as night flying versus day flying), cannot be compared with your flight. We shall then want to know the following: On how many of these flights did an accident take place? On those flights in which there was an accident, what percentage of the passengers were killed and what percentage were injured? By combining these figures according to the laws of chance, we obtain your risk. For example, we might find from the records of comparable flights that there was an accident in 1 flight in every 34,000. The chance that you will be involved in an accident is therefore 1 in 34,000. Suppose, further, that the records show that when an accident does occur, on the average seventeen passengers out of twenty are killed, two out of twenty injured, and one out of twenty is safe. Then your risks are as follows: Chance of being killed, 1 in 40,000; chance of being injured, 1 in 340,000; chance of safe arrival, 679,981 out of 680,000.

In finding your risk on your trip from New York to Washington, we have used neither passenger miles nor passenger hours. Rather we have used fatalities or accidents per trip. One obvious reason for the inadequacy of both passenger miles and passenger hours is that the risk is not spread out equally over miles flown or hours flown. It is concentrated above all in the take-offs and landings. The risk in a nonstop flight from New York to Chicago is certainly less than in a flight in which the plane puts down at several intermediate points.

On the other hand, suppose that you are at an airport and have decided to cruise around in a plane for an hour or two, just for the fun of the thing. Your risk is made up of three components: the take-off risk, the flight risk, and the landing risk, and no two of them are the same. The first has nothing to do with length of flight, while the second depends directly on length of flight. The third, like the first, is independent of it, barring unusual circumstances. In this last example, nothing could be more absurd than to measure the risk in terms of passenger miles, since miles are invariably measured from where the ship takes off to where it puts down. Here the distance would be at most a few hundred feet, which would lead to prodigious estimates of risk.

The correct measure depends on the circumstances, and even on the purpose of the flight. This indicates how difficult a comparison between the safety of air travel and of other methods of transportation really is. If a man in New York must get to Washington and has his choice of travel by air, by rail, or by bus, neither the distance between the points nor the time of transit is a true measure of the risk. But since distance between two given points is very roughly the same by air as by rail or bus, the rate of fatalities or accidents per passenger mile comes closer to furnishing a single fair *comparative* basis than any other.

Statistics on railroad fatalities in the United States furnish an interesting and simple example of the importance of careful analysis of the precise meaning of all data used. During the year 1946 the number of deaths chargeable to steam railroads was 4,712. At first glance this seems like a large number of fatalities and one that is not wholly consistent with the safety records claimed by the railroads in their advertising. If this figure is analyzed, however, we find that of the 4,712 only 132 were passengers. There were 736 railroad employees, and 1,618 trespassers, mostly riding the rods. In addition, 2,025 people were killed at grade crossings. Under the heading "other non-trespassers" there are listed 201 fatalities. This analysis of the figures (given out by the National Safety Council) is clear provided that we understand the meaning of the terms used. For example, passenger fatalities include not only those traveling on trains, and those getting on or off trains, but "other persons lawfully on railway premises."

All the examples discussed in the last few pages indicate the necessity for careful and skillful analysis of data before attempting to draw conclusions, if fallacies are to be avoided. A further notorious example of the sort is the statistics on automobile accidents, especially as applied to large cities. Accident figures for a city are usually given in the form of number of accidents of various types per 100,000 of population. Another, and for some purposes more effective way of presenting these figures is number of accidents per 100,000 passenger cars. Traffic engineers have studied this problem and have set up a large number of significant categories, including time of day, age and condition of car, type of accident (two cars, car and truck, car and pedestrian, and so forth), condition of surface of road, experience and physical condition of driver, and many others. From skillfully made analyses of this sort it is possible to determine, with a minimum of guesswork, what steps would be required to reduce materially the number of automobile accidents. There appear to be few that are both effective and practicable.

There are certain widespread beliefs, that come under the head of statistics, and that do not appear to have been reached by sound statistical reasoning. An example is the statement, a favorite theme of some popular writers, "Greatness in men is not, as a rule, transmitted to the next generation." This is maintained in spite of such records as those of the Bach, the

Bernoulli, the Darwin families, and some others. The fallacy here, and it almost certainly is one, is in comparing the son of a great father with the father himself. Clearly the comparison should be made with a random sample of individuals. Furthermore, greatness is scarcely a proper basis for comparison between generations, because it contains a large element of accident. Talent or ability would serve the purpose better. On this subject it may also be observed that inherited characteristics are transmitted through both the father and the mother. For some strange reason, the mother is usually left altogether out of these discussions. Even so, the intelligence of sons shows a high correlation with the intelligence of fathers, just as the height of sons shows a high correlation with the height of fathers.

Very similar in character is the popular belief that infant prodigies usually fade out before reaching maturity. This is an excellent example of the effect on people's thinking of a biased sample. In this case the biased sample consists in the wide publicity given to a few boy prodigies who graduated from college at an early age, and who afterward were discovered in some unimportant position. Such cases are the only ones that register in the public mind. An adequate analysis of the problem demands a comparison between the careers of a group of child prodigies and a group representing a random sample of children of the same age.

In certain specialized fields, such as music, it is not necessary to make an elaborate survey to determine the probable future status of child prodigies. Reference to biographies of leading musicians discloses that, almost without exception, they were infant prodigies. For example, it would be difficult to name a first-class concert violinist who did not play important works in public before his tenth birthday. In order to complete such an analysis, it would be necessary to find out the number of musical prodigies that did not become top-flight concert artists. We are less interested here, however, in the truth or falsity of this popular belief than in the statistical fallacies on which it rests. Many correct conclusions, especially in matters where there are but two possibilities, are based on utterly erroneous reasoning. If you toss a coin to decide the question in such a case, your chance of being right is 1 in 2.

A professor of psychology once conducted a frequently repeated experiment on the subject of evidence. During class he engaged several people to enter the classroom suddenly, start a near riot, and go through certain other prearranged motions. When the disturbance had quieted down, the professor asked each member of the class to record his impressions as accurately as he was able, including such things as how many individuals entered the room, exactly what each did, and so on. When the results were tabulated it was found, rather astonishingly, that only some 10 per cent of the class had indicated correctly the number of individuals that were involved, not to mention the more detailed facts.

As far as I know, this experiment has always been interpreted on the basis that those students who correctly wrote the number of intruders really *knew* this number. We submit that on the basis of the laws of chance it has not been proved that any of the students *knew* the number of intruders. From the account of the experiment it seems likely that if the students had been blindfolded and merely knew that there had been a few intruders, something like the same number would have guessed the correct answer. It would indeed be interesting to have a "frequency table" of these answers, giving the number of students who said there was one intruder, the number who said there were two intruders, and so on. We are not suggesting, however, that this reasoning be applied too seriously to the rules of evidence in court, to which the professor's experiment has obvious applications.

Possibly the most frequent cause of fallacious statistical conclusions is the use of a *biased sample*. We have already discussed briefly in Chapter XIV the connotation of this term. It means a sample which fails to reflect correctly the essential characteristics of the entire population from which it was taken. The first critical question is the size of the sample. Suppose, just before a presidential election, you ask one hundred voters how they intend to vote. Whatever the result, it would be completely meaningless; for presidential elections are usually so close that a relatively small shift in percentages would swing the election one way or the other. But if you ask one hundred voters whether they would favor the return of a national prohibition law, the result would be so one-sided that it would require an extremely biased sample, even of one hundred people, to give the wrong answer. So we see that the size of the required sample depends not only on the size of the population, but also on the type of question to be answered.

Even though the sample is a very large one, it does not follow that it adequately represents the population as a whole. For an example of a large but seriously biased sample we cannot do better than go back to a poll of voters in the presidential election of November 2, 1936, conducted by the *Literary Digest* magazine, referred to briefly as the *Digest* poll. This was by far the largest of the several political polls, ballots being mailed to no less than ten million voters, with over two million responses. This magazine had been making similar polls for a quarter of a century and had always correctly indicated the result of the election. So when the poll showed, and continued to show, a plurality in favor of Landon against Roosevelt, it was a matter of widespread interest and great astonishment, for all other available information, including that furnished by the other polls, indicated precisely the opposite. As the fateful election day approached speculation reached a high pitch. People all over the country were asking each other, "Can the *Digest* poll possibly be right?"

Now this is the famous election in which Roosevelt carried every state except Maine and Vermont. Of the total vote of

45,647,000, Roosevelt got 27,752,000. So to say that something was wrong with the *Digest* poll would be putting it rather mildly. What is interesting is to find out where this bias lay.

First of all, let us put ourselves back to the end of October, 1936, with the completed *Digest* poll before us and the election still some days off. The fact is that the poll itself contained convincing evidence that something was radically wrong, and this evidence was seen and interpreted at the time by many competent statisticians. The nature of this evidence is as follows: The completed poll carried figures showing "How the same voters voted in the 1932 election." These figures showed that 1,078,012 of those who sent their ballots to the *Digest* poll voted the Republican ticket in 1932, and 1,020,010 the Democratic ticket. But this is definite proof of a biased sample, for it actually shows more Republican 1932 voters than Democratic, while the 1932 election itself showed 22,822,000 Democratic votes against 15,762,000 Republican votes.

In other words, in 1932 there were about 45 per cent more Democratic than Republican voters, and in a fair sample of 1936 voters that same proportion should have been maintained among those who voted in 1932. The poll also indicated the number of voters that had shifted from one party to another between 1932 and 1936. But in view of the major bias just pointed out, it was necessary to assume that a similar bias affected these figures. This means that it was precarious, statistically speaking, to attempt to correct the error in the poll numerically, but the crudest correction was sufficient to indicate that Roosevelt was running ahead. As a matter of fact, it was not possible, based on these 1932 figures, to make a correction swinging enough votes over to Roosevelt to indicate anything like the actual majority.

It is obvious that the *Digest* sample was biased in several respects. What were these other biases? There is no conclusive evidence about them, but on the basis of the facts we can reach one or two conclusions that are probably correct. It is significant that the other polls, like the Gallup Survey, which depend for their success on smaller samples selected with great care and skill, gave the election to Roosevelt. But it is also significant that no one of these polls predicted anything like the landslide that took place. Each was considerably in error, so that if the election had been close, it would have picked the wrong man as winner. The probable explanation is that there was a real shift in opinion during the last weeks of the campaign, possibly even during the last days, not sufficient to change the general result of the election, but enough to change the figures materially.

These same considerations apply with even greater force to the election of November 2, 1948, in which every poll erroneously gave the election to Dewey. So much has been written about this failure of the polls that we shall content ourselves here with only a few comments.

Opinion polls are statistical predictions based on small samples. Such predictions necessarily involve error, and the smaller the sample, other things being equal, the larger the error. The polls are therefore wrong, to some extent, on every election, and in the case of a very close election, some of them are almost certain to pick the wrong candidate. But in November, 1948, all the polls predicted that Truman would lose; all the errors were in the same direction. This fact indicates clearly that there was a common source of error, and there is much evidence as to its nature. All the polls showed that during the campaign there was a marked swing toward Truman, just as there was toward Roosevelt in the election of 1932. One of the polls, in fact, stopped interviewing voters months before the election, convinced that Dewey's large lead could not be overcome. There can be little doubt that this shift of potential votes from one candidate to the other during the campaign was a major reason for the failure of the polls.

There has been much speculation as to whether this shift came from the "undecided" column or from the "Dewey" column. This question cannot be settled and is not of great importance in any case. For many voters simply do not know how they will vote until they arrive at the voting booth. When asked how they intend to vote, they indicate one of the candidates or say they are undecided, depending on their temperaments and habits of thought. This problem is one that the polls have not successfully solved.

It is safe to say that no polling technique based on small samples will ever predict with accuracy a close election, nor will it predict with accuracy a fairly close election in which there is a rapid shift in sentiment just before election day.

This question of rapidly changing conditions indicates another source of fallacy in statistics, namely the time element. It is of the greatest importance in economics and in government. In many instances, as for example in the case of unemployment, by the time the statistics have been collected and analyzed, the situation itself may have changed completely.

The biased sample is a potential source of error in surveys of various varieties, especially in those that are mailed. The replies are seldom received from more than a small percentage, and the sample is subject to the serious bias that there are large classes of people who never or practically never respond to such appeals. In some cases the sample may be of value in spite of this defect, but usually in such cases the conclusions obtained are obvious without a questionnaire, and the results should be classified as propaganda rather than as evidence.

Many of these difficulties can be avoided by using the personal-interview technique. In fact, when the precise composition of the sample is a matter of some delicacy and when this composition is vital to the success of the survey, the interview technique is almost indispensable. Of course, the interviews must be skillfully conducted, the questions properly prepared, and the results subjected to thorough statistical scrutiny. Even so, there are many pitfalls in surveys of this type, and I have seen many that can be classified only as silly. In recent years it has, in fact, become common practice to attempt to find the answers to puzzling questions, in the social sciences, in administration, in business, by appealing to the public. Each individual interviewed may also be puzzled by these problems, but being pushed to say something, he follows the path of least resistance and indicates one of the listed answers as his own. The idea seems to be that even if no individual knows the answer, if you ask enough individuals the majority opinion will somehow be the right one. Such instances are merely abuses of a method of getting information that can be basically sound and has yielded valuable results not otherwise obtainable. Unfortunately, no technique is foolproof. Those involving statistics seem to be the least so.

CHAPTER XVII

Statistics at Work

IT IS perhaps time to pause in our survey of statistics and its uses long enough to see where we are. In Chapter XIV we attempted to make clear the nature of the modern science of statistics and, with a few simple examples, to indicate the type of problem to which it is applicable. In the next chapter we discussed and illustrated some of its basic ideas, such as frequency distributions, various sorts of averages including the standard deviation, and the notion of correlation. Finally, in Chapter XVI we tried to make these fundamental notions sharper and clearer by pointing out misuses of them, unfortunately all too common, that lead to fallacious statistics.

Having discussed in an elementary way the root ideas of statistics, we are now ready to see how they are put to work in a wide variety of fields. We shall be able to cover only a small part of the ground, however, as statistics in recent years has spread itself like a drop of oil on a sheet of water. We shall confine our illustrations to fields that are, we hope, at least fairly familiar and of some interest to the general reader. We have had to restrict ourselves further to examples whose interest is not dependent on more technical or mathematical know-how than is here assumed.

Let us begin with an illustration from the field of sports. The first world's record for the mile run was established in England in 1865. It was 4^m $44^s.3$, or more briefly 4:44.3. As we write this (1948) the record stands at 4:01.4, set by Hagg of Sweden in 1945. In eighty years the record has been lowered by almost three quarters of a minute. How much lower can it go? Eighty years is not a long interval, in terms of the future, and it seems reasonable to expect continued improvement. One magazine writer even expresses the view that someone will always break the record, no matter how low it goes. This is of course a large order. We shall now see whether statistical methods can throw any light on this question. The first step is to study the data. Table XXVIII gives the history of the mile run (we are referring to outdoor races only). A glance at it discloses several in-

TABLE XXVIII

Year	Athlete	Country	Time
1865	Webster	England	4:44.3
1866	Lawes	England	4:39.0
1868	Chinnery	England	4:33.2
1871	Chinnery	England	4:31.8
1874	Slade	England	4:24.5
1881	George	England	4:19.8
1884	George	England	4:18.4
1895	Bacon	England	4:17.0
1895	Coneff	U.S.A.	4:15.6
1911	Jones	U.S.A.	4:15.4
1913	Jones	U.S.A.	4:14.4
1915	Taber	U.S.A.	4:12.6
1923	Nurmi	Finland	4:10.4
1931	Ladoumegue	France	4:09.2
1933	Lovelock	New Zealand	4:07.6
1934	Cunningham	U.S.A.	4:06.8
1937	Wooderson	England	4:06.4
1942	Hagg	Sweden	4:06.2
1942	Hagg	Sweden	4:04.6
1943	Andersson	Sweden	4:02.6
1944	Andersson	Sweden	4:01.6
1945	Hagg	Sweden	4:01.4

teresting things. We notice, for example, that during the first ten years of competition the record dropped some twenty seconds, almost as much as during the succeeding seventy years. This is typical of a new competitive sport. We can break the table down along national lines. The first thirty years might be called the English period, the next twenty the United States period, the next twenty-five the international period, and the final five years the Swedish period. This last may well have ended, since Hagg and Andersson were declared professionals in 1946. We notice also the blank spaces in the table. During the first decade of the century, for example, there was no new record, and there was only one between the years 1915 and 1931. On the other hand, during the thirties the record changed four times, and in the first half of the next decade it changed no less than five times.

In a thorough statistical study every one of these special features of the table would be noted and interpreted. This would require an intimate knowledge of the history of the sport, including such matters as the various styles and schools of running and the art of applying them in competition. This would lead to adjustments in the table and quite likely to breaking it down into two or more periods.

We have not gone into these essential refinements but have assumed that the irregularities have averaged out sufficiently to permit an over-all conclusion. We find by statistical analysis typical of that employed in many studies that Table XXVIII leads us to the conclusion that someday in the future the record could go as low as 3:44. This means that if the period 1865 to 1945 turns out to be representative of the future in every sense, then it is probable that the record will continue to be beaten until some such mark as 3:44 is reached.

The method used in problems of this sort can be most profitably discussed from the standpoint of common sense in statistics. We have reached a conclusion as to how the mile record was changing during the eighty-year period covered by the data, and have then projected it into the future. This is technically known as *extrapolation*. Immense care and sound judgment are required to avoid reaching conclusions in this way that are nothing short of ridiculous. A simple illustration will show what sort of danger exists; but quite often the traps which nature sets for the statistician are much more subtle. Suppose that we had, in this very problem, extended our results *backward* in time, instead of forward. Then the same formula that tells us that the ultimate future record is 3:44 tells us that the record, if there had been one in those days, would have been 3,600 seconds, or one mile per hour, in 1977 B.C.

Does this mean that our figure of 3:44 is quite absurd? Of course not. The fallacy is easily pointed out. Whenever men start doing something new, such as running a competitive mile, heaving a weight, driving a golf ball, or knocking home runs, it takes some little time to get the hang of the thing. Just what year they start doing it is quite unimportant. You can be sure that if the Egyptians of 1977 B.C. had taken it into their heads to see how fast they could run, their hourglasses would have indicated something comparable to the 1865 record.

Since we have got onto the subject of track athletics, there is one other little problem that we might point out. At the start of a short race (one of the dashes) the runner must develop full speed as fast as he can. He necessarily loses a little time, as compared with a running start. Can we determine statistically how much time he loses? In Table XXIX we give the records for the 100-, 220- and 440-yard distances, together with the corresponding speeds in feet per second. The speed

TABLE XXIX

Distance in Yards	Time in Seconds	Speed in Feet per Second
100	9.4	31.9
220	20.3	32.5
440	46.4	28.5

for the 220 is greater than that for the 100. It is clear that this situation is caused by the loss of time in starting, for otherwise we would expect the speed for the longer race to be the same or somewhat less, certainly not greater. Let us make the assumption that the effects of fatigue do not show themselves in as short a race as the 220. Then, apart from the effect of the loss of time in starting, the 100 and the 220 are run at the same speed. Now the time loss in starting is the same for the two races; it lowers the average speed more in the 100 because the race is shorter, so that there is less time to pull up the average. Under this assumption we can compute the loss of time in starting by elementary algebra. It is 0.3 seconds. If we assumed that the 220 is run slightly slower than the 100, due to fatigue, we would get a result slightly greater than 0.3 seconds. The table tells us that in the case of the 440 the effect of fatigue is pronounced. But let us stick to our figure of 0.3 seconds. If this lost time is subtracted from the time for the 100, we get 9.1. This is what the record should be for a running start. It corresponds to a speed of 33.0 feet per second. The same speed would of course apply to the 220, with a running start. This speed of 33 feet per second must be very close indeed to the fastest that man has ever traveled on a flat surface entirely on his own.

We wish to call attention to one feature of this little problem that is often of importance in applied statistics. By expressing the records for the dashes in terms of feet per second we were able to see certain relations at a glance that are not equally obvious when the records are expressed in their usual form. Nothing new was added; only the form was changed. In other words, in Table XXIX the first and second columns alone say exactly the same thing as do the first and third alone. Each set implies the other. This is the purely logical approach. But in science, and therefore in applied statistics, the major problem is to find relations between things, and in this job the role of intuition is paramount. It follows that if the data are expressed in such a way—it will always be logically equivalent to the way the original data are expressed—that our intuitions are brought into play, we shall go further with a fraction of the effort.

This is nowhere more evident than in the use of charts and graphs. Complex data can in this way be absorbed and understood in a small fraction of the time otherwise required. In some cases it is even doubtful that the same understanding could ever be reached without this visual aid. Take the case, for example, of an ordinary topographic survey map of a region. Logically, this map is strictly equivalent to the hundreds of pages of measurements and notes representing the work of the surveyors. But most people in five minutes would learn more from the map about the geography of the country, its lakes, rivers, hills, valleys, roads, and other features, than in five hours, or even five days, from the surveyors' data. And it is a very rare person who could ever succeed in visualizing the country from the latter as effectively as from the former.

This matter of drawing up the data in such a way as to lead to a quick intuitive grasp of the essentials is of particular importance in business, where complex numerical data must often be presented to executives who need the information but can spare little time to get it. That the importance of graphical presentation of such complexes of facts is not always appreciated, even by those with scientific training, was brought home to me once when an industrial chemist said, "What is the sense of drawing up these charts, anyway? After you get done you haven't added anything new—it was all in the original data." The most charitable interpretation of this remark is that it was aimed at a small but well-defined group called *chartmaniacs* whose passion for charts is so intense that they believe even the simplest data to be incomprehensible unless presented in this manner.

As a simple illustration of the fertility of charts we have added Table XXX, with the suggestion that the interested reader take the trouble to put it in graphical form. He will find

· 275

the resulting chart informative. The data are simply an extension of the first and third columns of Table XXIX to races as long as 5,000 meters. We have omitted several of the metric distances that are practically duplications of those in the English system (yards). They can easily be added, if so desired. The figures in the table show an amazing regularity, as disclosed by a chart. They are an example of a complex statistical law at work. It is interesting to speculate on the meaning of the limiting speed to which the values seem to tend as the length of the race is increased. The record for the marathon run of slightly less than $26\frac{1}{4}$ miles (46,145 yards) is 15.5 feet per second. The record for the two-mile walk is 10.8 feet per second.

TABLE XXX

Distance		Speed in Feet per Second
Yards	*Meters*	
100		31.9
220		32.5
440		28.5
880		24.2
(1,093.6)	1,000	23.2
(1,640.4)	1,500	22.1
1,760	(one mile)	21.9
(2,187.2)	2,000	21.0
(3,280.8)	3,000	20.5
3,520	(two miles)	20.2
5,280	(three miles)	19.5
(5,468.0)	5,000	19.6

During World War II there developed a novel and interesting application of the scientific method to military problems, which came to be known as operations research, or operational research. It is, in essence, the scientific study of a large and complex organization, with a view to increasing its efficiency and effectiveness. Its methods and results are therefore applicable to any complex organization, for example a business, and it is in fact equivalent to business research, in the best sense of that often misused phrase. While, like all scientific research, it uses ideas and methods that seem useful, regardless of their source, it inevitably involves large amounts of probability theory and statistics. The military results were of course secret during the war, but some few have since been made public.

Operations research developed out of the Battle of Britain. Its first problem was to determine whether the radar interception system was being used to the fullest advantage. From this simple beginning it developed rapidly, due to its striking successes in the field, until later in the war operations research was at work with practically every important command in the British armed forces. After the United States entered the war, similar units were set up in this country. This was due largely to the influence of Dr. Conant, then president of Harvard, who was in England during the Battle of Britain and who is reported to have expressed the view that the contributions of the operations research groups were of major importance in winning that battle. The American units were organized beginning in the spring of 1942 and were ultimately attached to all the major services.

One of the early leaders in operations research was P. M. S. Blackett, professor of physics at Manchester University, and Nobel prize winner in physics for 1948. Professor Blackett has said of the nature of the work that it was "research having a strictly practical character. Its object was to assist the finding of means to improve the efficiency of war operations already in progress, or planned for the immediate future. To do this, past operations were studied to determine the facts; theories were elaborated to explain the fact (and this is very important; operations research is not merely the collection of statistics about operations); and finally the facts and theories were used to make predictions about future operations and to examine the effects of various alternatives for carrying them out. This procedure insures that the maximum possible use is made of all past experience." This statement, with the change of a few words here and there, is an excellent description of practical research in any field. The only difference between practical research and theoretical research is that in the former the primary object is to learn how to do things better, while in the latter the primary object is to know things better.

We turn now to an example of operations research in action.

When a bombing plane spots a hostile submarine a certain time elapses before the plane can reach the point of attack and drop its depth charges. The submarine will submerge the instant that its lookouts, acoustic, optical, or radar, give warning that the aircraft is approaching. If the submergence time is half a minute, a plane flying 180 miles per hour will have covered a mile and a half during this interval. Records were kept showing, for a large number of attacks, how long the submarine had been submerged at the instant that the plane passed over the point of attack. One series of such records is shown in Table XXXI. The percentages in the table represent,

TABLE XXXI

Time Submerged in Seconds	Percentage Submarines Attacked
Not submerged	34
0–15	27
15–30	15
30–60	12
over 60	11
	100%

of course, averages. If there were much scattering, or dispersion, about these averages, the table would be worthless. This was not the case, and furthermore, over a long period of time and over a wide area of ocean these percentages were approximately the same.

Even if it were assumed that the depth charge is fused for exactly the right depth, the probability of success of the attack decreases rapidly with the length of time the submarine has been submerged. In the first place, the location of the submarine must be estimated in terms of the point at which it submerged, in other words by the rapidly dispersing swirl left behind by the submarine. In the second place, the direction of the submarine's course is not known. It is evident that any means whereby this interval of time could be decreased would greatly increase the percentage of successes. Taking account also of the uncertainty in estimating the depth, the advantage of such measures is even more obvious. The problem, then, can be stated in this way: to find measures that will significantly increase the probability that the plane will detect the submarine, while at the same time the probability is small that the submarine will detect the plane at distances much over one and a half miles. It turned out that for a 20 per cent decrease in the distance at which the airplane could be detected by the submarine, there was a 30 per cent increase in destruction of submarines. This shows the importance of even small increases in air speed and in effectiveness of camouflage. This work led to large improvements in technique.

Another illustration having to do with submarines is the following: Losses to submarines of convoyed merchant ships in the North Atlantic crossing were carefully studied, and a statistical relation between the relative losses, the number of ships, and the number of convoy vessels was found. If we call r the ratio of ships lost to total ships, N the number of ships, and C the number of convoy vessels, this relation can be written

$$r = \frac{k}{NC} \, ,$$

where k is a constant whose numerical value was determined by experience. It is clear that the ratio of losses r decreases if either N or C is increased. For example, suppose that N is 50, C is 10, and that k has been found to have the value 10. Then

$$r = \frac{10}{(50 \times 10)} = 2 \text{ per cent.}$$

Now suppose we increase the number N of ships convoyed from 50 to 100, leaving C unchanged. The equation then gives $r = 1$ per cent. The number of ships lost is the same in each case, since 2 per cent of 50 is equal to 1 per cent of 100, each being equal to 1. This formula led immediately to a most important conclusion, the application of which saved many ships and many lives. If convoys were sent out only half as often, there would be twice as many escort vessels available. Now according to the equation

$$r = \frac{k}{NC},$$

if you double the number of ships and at the same time double the number of convoy vessels, you cut in two the number of losses. Just as many ships set sail as under the previous plan, but only half as many are lost, on the average. This is an example of a really radical improvement in operations as a result of a statistical investigation.

Among other outstanding examples of operations research that have been published are the following: How many planes should be used in a bombing raid, given the nature of the probable defenses? How much time should the personnel of bombing squadrons devote to training, as compared to operational missions, in order to deliver the maximum bomb load on the target? We shall content ourselves with mentioning these problems and turn to other applications of the statistical method.*

* The reader interested in further details is referred to the following books: S. E. Morison, *The Battle of the Atlantic,* Little Brown & Company, Boston, 1947. J. P. Baxter 3rd, *Scientists Against Time,* Little Brown & Company, Boston, 1947. L. R. Thiesmeyer and J. E. Burchard, *Combat Scientists,* Little Brown & Company, Boston.

The study of populations provides a good example of statistics at work. If you draw a geographical line around a city, a community, or a country, and determine its present population by taking a census, its population at any future time then depends on four factors. They are births, deaths, immigrations, and emigrations. They take account of processes of creation and destruction within the geographical area and of passages of human beings across its boundary line in either direction. In order to predict the future population of the area it is necessary to project ahead the rates of births, deaths, immigrations, and emigrations, which are continually changing, sometimes quite abruptly. The difficulties and uncertainties of this task are very great.

The birth rate varies considerably from one region to another, depending on the ideas and attitudes of the population, for example their religious beliefs. It depends on biological and social factors. It is indirectly but forcibly affected by economic conditions, present and anticipated. Following a major war there usually takes place in each belligerent country, regardless of economic conditions, a sharp increase in the birth rate. A glance at the statistics published by the Statistical Office of the United Nations discloses how pronounced this effect has been after World War II. In most cases there has been so large an increase that temporarily, at least, the birth rate far exceeds its prewar level.

The death rate is obviously affected by wars, famines, pestilences, floods, earthquakes, and other natural disasters. Owing to sanitary improvements and medical advances it has de-

---

The following articles can also be consulted: C. Kittel, "The Nature and Development of Operations Research," *Science*, February 7, 1947; J. Steinhardt, "The Role of Operations Research in the Navy," *United States Naval Institute Proceedings*, May, 1946.

I wish to acknowledge my indebtedness, in connection with the brief account in the text, to a lecture by Dr. Steinhardt on "Operations Research," delivered before the Institute of Mathematical Statistics and the American Statistical Association on Dec. 29, 1947.

creased steadily over the years. As a result the average length of life has increased from somewhat less than thirty years in the England of Queen Elizabeth to over sixty years in the England of today. The greater part of this change has taken place during the past century and is due largely to the enormous decrease in infant mortality. At the present time the death rates for infants under one year of age vary greatly from one country to another, depending on conditions in the particular country. A rough measure of these rates is determined by the change in life expectancy at birth and the life expectancy at age one. A table of world life expectancies* discloses the following facts: In Australia the life expectancy of a male at birth is 63.5 years, at age one 65.5. The life expectancy of the child therefore increases two years, on the average, if he survives the first year. In Hungary, on the other hand, the life expectancy of a male at birth is only 48.3 years, while at age one it jumps to 57.1. The life expectancy increases almost nine years, therefore, if the child survives the first year. This indicates a very high infant mortality compared with that of Australia. Hungary shows a lower life expectancy than Australia for every age group, but the disparity is not so serious after the first year.

Immigration and emigration are even more erratic and unpredictable than rates of birth and death. Great movements of population have taken place in the past, under the pressure of overpopulation and the lure of new and underpopulated lands, but today such movements are smaller and even less predictable. They are largely caused by revolutions and persecutions, and are affected by immigration laws, even by emigration laws.

Summing up, it is clear that any projection into the future of the four factors controlling population growth, for more than a few years at a time, is extremely precarious. It follows that estimates of future population are unreliable. We can get rid of two of the factors, namely immigration and emigration,

---

* For example, in *Information Please Almanac* for 1948, page 284.

by extending the geographic area to include the entire earth. However, since every person is born and dies, while few migrate, this simplification is not too helpful.

The birth and death rates that we have been talking about are what are known as the *crude* rates. They are obtained by dividing the actual number of births and deaths for the year by the mean population, the latter being expressed in thousands. In a community with a population of 10,000 and 200 births during the year, the crude birth rate is 200/10, or 20 per thousand of population. The use of the entire population as a base is not entirely satisfactory for certain types of population studies, and various modifications have been introduced. One such refinement is to use as base the number of married women in the childbearing age brackets, usually taken as from fifteen to forty-five. In this way births are related to that section of the population most directly responsible for them.

In the case of death rates the situation is somewhat different. From the simple and fundamental fact that every person dies sooner or later it follows that refinements of the death rate take the form of relating the number of deaths in each age group to the number of individuals in that group. This leads us back to the notion of life expectancy, which we have already considered.

If the number of births exceeds the number of deaths, the population is increasing. (We are leaving immigrations and emigrations out of account.) This means that the base, against which both birth and death rates are measured, is itself increasing, so that these rates are distorted. In the opposite case, with a decreasing population, there is a corresponding distortion. Without going into the mathematics of the matter, it can be seen how this works out by setting up an imaginary community on paper. To avoid tedious complications some rather queer assumptions will need to be made, but the principles will not be affected. Start with a population of 1,000, for example, and assume that for several years births and deaths are each equal to 150, so that the population is stationary. This will get the study started without complications. Now increase the birth rate to, say, 200, and make the assumption that every individual dies exactly five years after he is born. You now have an increasing population which can be computed for each year merely by adding to the figure for the previous year the number of births and subtracting the number of deaths. Knowing the population and the number of deaths for each year, you can determine the death rate by dividing the latter by the former and multiplying by 1,000. In this way you can see what happens to the series of death rates under the circumstances set up. Other types of situations, for example a declining population, can be studied by changing the basic figures that were assumed. If we make the convention that each year in the above scheme represents ten actual years, we will have a rough, simplified model of a human population.

The interest and importance of population studies were first realized as a result of the work of Malthus at the beginning of the nineteenth century. He noted the tendency of all living things to multiply in geometric ratio. For example, a single cell divides into two new cells, each of these giving rise to two more, and so on; in this case the ratio is two. From what he could learn Malthus concluded that the human race tends to double its numbers every twenty-five years. On this basis a country with a population of ten million in the year 1800 would by now have a population of nearly six hundred and forty million, which is almost one third the population of the globe. This led him to a study of the basic reasons why this tendency to increase is held in check. The first and most obvious reason is the supply of food, using this word to represent all the necessities of life. For on the most liberal estimate, taking account of past and probable future scientific improvements, it is not conceivable that the food supply could be long increased at the rate of doubling itself every twenty-five years. And if it is increased at a slower rate, either the amount of food available per individual must sink lower and lower, or population growth must somehow adjust itself to the slower rate. This cramping effect of food supply on population growth is an example of what Malthus called "positive checks." Other examples are wars and epidemics. As an alternative to these Malthus discussed what he called "preventive checks." They consist in lowering the birth rate when overpopulation threatens, rather than increasing the death rate. The operation of the positive checks means destruction, suffering, hunger. That of the preventive checks means coordinated planning for a better and fuller life. It appears that the question of future overpopulation is one of great concern at this time, and an international committee, sponsored by the United Nations, is at work on it.

We shall close this brief review of statistics in action with a final example, taken from one of the standard elementary texts.* It is clear that there must be a relation of some sort between the prices of industrial common stocks and business activity, since stock prices are related to earnings and prospects of earnings of industrial companies, and these earnings, by and large, are dependent on business activity. The problem studied was this: Do changes in the price level of stocks foreshadow changes in the level of business activity, or is it the other way around? Whichever way it is, what is the time lag involved? Two periods were examined; the first covered the eleven years immediately preceding the outbreak of World War I, the second the years 1919 to 1937. In each case index numbers representing the prices of industrial stocks were compared with index numbers representing business activity. It was found that in the earlier period stock price changes anticipated changes in the activity index by from three to seven months, the average being five. However, this average was not sharply

* F. C. Mills, *Statistical Methods*, Henry Holt and Company, Incorporated, New York, 1938.

defined, indicating a good deal of dispersion, so that any figure from 3 to 7 would serve satisfactorily. In the later period it was found that stock price changes anticipated changes in business activity by one month, on the average. Although this average also is not very sharp, the probability is high that the different results in the two periods correspond to a real change in conditions. The important thing is that in each case it is the level of stock prices that changes first. If it goes up or down, it is a good bet that within a month or two business activity will do likewise, provided that these results remain valid.

It is interesting to speculate on what would happen to speculation if this study, or an equivalent one using some other index than that of business activity, had come out the other way around. In that case the rise or fall of the index would announce in advance a corresponding rise or fall in the general level of stocks. Of course it would announce it only the way a statistical law does, as a probability. Furthermore, the probability that any individual stock would respond to the signal would be much smaller than that for the composite stock index used in the study. Nevertheless, the effect on speculation would be considerable. For this information as to the probable future course of the market would tend to narrow the differences of opinion on which speculation thrives. The general effect would be to reduce speculative activity. This reasoning is probably not applicable to the widely held Dow theory of movements of common stocks. The latter is an empirical theory based on the behavior of the industrial and railroad averages themselves, without reference to any economic index. The effect of the Dow theory is probably small not only because it is by no means universally accepted by traders, but more especially because it gives signals only under special conditions, which by their nature are relatively rare.

Speculation and statistical prediction are compatible as long as the latter is not too accurate. At this writing there appears to be little danger that speculation will die out for this reason.

For the possible fields of speculation are almost limitless, and if statistics should catch up with one of them, it would immediately be replaced by another well beyond the reach of statistics or of science. Only a few hundred years ago men could speculate on the date of the next eclipse of the sun; today this would make an exceedingly poor bet.

CHAPTER XVIII

Advertising and Statistics

THE problem of advertising, stated in a minimum of words, is to produce a maximum effect on a maximum number of individuals at a minimum cost. Advertising of a product is rated as successful if it tends, in the long run, to increase the net profits of the concern which handles the product and pays the advertising bills. Since past experience goes to show that advertising of certain products in certain manners can be extremely profitable, almost every business is faced with an advertising problem of greater or less dimensions.

By its very nature advertising deals with large masses of people (the larger the better, in a sense); so it inevitably involves problems in which statistical thinking is called for. In many of these problems the best route to a reliable solution is that of statistics, and to this end its entire modern technical machinery can be set in motion. In others precise statistical methods have only an indirect application, as we shall see, and in still others they have no application at all.

In all these cases, whether the advertising medium be newspaper, magazine, billboards, handbills, mailed circulars, radio, or some other, the advertiser must think in terms of the *statistical class* or *group* of prospects reached and attempt to estimate or measure its value as potential consumers of his product. To take an obvious instance, if he is advertising farm implements, he will naturally turn to farm journals. If he is advertising a new bond issue, he may place his ad on the financial pages of certain newspapers, or in a financial journal, where it will be seen by a comparatively small number of people comprising, however, a very select class, from the point of view of the advertiser. On the other hand, the manufacturer of farm implements, we may be sure, will not consider the readers of financial journals as favorable prospects.

In practice, the case is seldom so simple. The group of individuals exposed to any given type of advertising is usually composed of a large number of smaller groups, having various degrees of value as prospects. The statistical problem of the advertiser, or of the advertising expert, is to discover means of dissecting the entire group into these component groups, and to find means of measuring the value of each in relation to the advertiser's business. This value depends not only on the percentage of potential customers in the group, but in an equally important way on the amount of favorable attention that the advertising produces. If the advertiser's product is one that is used by all classes of people, the problem is largely one of sheer volume and of attention value. If the advertiser is offering products or services that appeal only to a very particular class of individuals, the statistical problem of locating and reaching this class is one of great importance. If, to take an obvious example, the advertiser is selling phonograph records, his prospects are contained in the group of phonograph owners. Waste circulation for his advertising will evidently be reduced to the lowest point attainable without the use of second sight, if he confines his distribution to this group. This he can do by use of a mailing list of names of individuals who have recently purchased phonographs. If he uses magazine advertising, his waste circulation will evidently be less in a magazine devoted to music than in others, and still less if the magazine contains a department devoted to phonograph records.

These self-evident statements are intended merely to indicate the type of analysis that is required in more complicated cases.

When the choice presents itself, as it very frequently does, between a small circulation containing a high percentage of true prospects and a large circulation containing a large number but a small percentage of true prospects, the decision involves a knowledge of the probable value of a prospect to the business, as well as the comparative cost figures. When reliable estimates of these figures are available, the matter can be settled by arithmetic. When accurate estimates are not available, which is more often the case, it is usually true that poor estimates are better than none, for they allow us to carry through our numerical thinking, and in many instances our conclusions will be the same as though it had been possible to use the correct figures, while in any case they are subject to all the control that other approaches to the problem can furnish.

The type of response that the advertiser is attempting to produce varies greatly from one case to another. It ranges from depositing a vague impression in an unoccupied corner of the prospect's mind, to inciting him to immediate action, clipping a coupon, writing for descriptive matter, or filling out a money order for a purchase. Evidently depositing vague impressions in people's minds is of value precisely to the extent that it leads, one way or another, to action on the part of someone. The preliminary impression may be merely one link in the chain that leads to ultimate action, but if without it the chain would be broken, it deserves its proportionate share of credit and must bear its proportionate burden of expense. In practice, of course, it is completely impossible to administer individual justice of this sort. Unless the action of the prospect is direct and immediate, and is stamped with an identifying mark of some sort, it is usually buried from view. The value of indirect and publicity or "name" advertising usually must be appraised, if it is appraised, by its long-term effects on the business of the advertiser. Of course there are occasional instances where indirect advertising produces an immediate and pronounced effect.

In advertising of the direct-action variety, aimed, let us say, at securing inquiries of some sort, the case is otherwise. As the inquiries are received a relatively short time after the dissemination of the ads, and as it is easy to arrange matters in such a way that each inquiry carries an identifiction of its source, the advertiser can obtain accurately the cost per inquiry for each medium of advertising. If he runs the same ad simultaneously in a dozen different mediums, the resulting figures, when chance fluctuations are taken into account, provide a sound basis for comparing the inquiry pull of one against another.

On the other hand, the advertiser may wish to test one type of ad against another by using the same medium more than once. The change may be in the printing, the arrangement, the terms, or in some other feature. Under these circumstances the advertiser is in the fortunate position of being able to determine by experiment whether a suggested change in his advertising produces more or less inquiries and can dispense with long and involved psychological arguments. But the number of inquiries received is not necessarily an adequate measure of their value. The word *value* here means precisely the net profit or the net loss that ultimately results from these inquiries. Each inquiry received must be followed up in an effort to convert it into a sale, and this follow-up entails a certain expense. In general, only a small percentage of inquiries are converted into sales. When such a conversion is made, it represents not only a sale, but the "birth" of a customer, who has a very real value to the business.

This value may in some instances be negative. If so, it means that the customer, or group of customers, entails more expense than the revenue produced and is therefore responsible for a net loss. The period of time used in measuring the value of a customer to a business is of the utmost importance. If a period of one year is considered, for example, the value of the customer may be negative, while if the period is extended to two years, the value becomes positive. If a period of several years is required for the value of the customer to change from negative to positive, few businesses would consider his acquisition a sound investment. In estimating this period it is evident that predictions of future economic conditions, among the classes of people involved, enter in a fundamental way. The accuracy of these estimates therefore depends on the accuracy of the economic predictions.

When the advertiser is in a position to know the advertising cost per inquiry, the net cost of converting an inquiry into a sale, and the value to his business of such a conversion, over a properly determined period, he is able to give a definite answer to the question, "What will be my net profit per advertising dollar in the medium of advertising I am considering?" These considerations are more directly applicable to the mail-order business than to any other, but in other cases the type of thinking required is the same.

When coupon or inquiry advertising is for the purpose of acquiring prospects for direct mail selling, it is possible, as indicated, to follow through all the way to the sale and to determine all the figures necessary for a complete appraisal of the value of the advertising. When, however, the selling is of the indirect variety, for example through dealers, it is not at all an easy task to determine how many actual sales should be credited to each hundred inquiries, or what the dollar value of such sales is. Under these circumstances the advertising expert is forced to make as good an estimate as he can of the value of responses received from a given medium of advertising—newspaper, magazine, radio, or whatever it may have been.

The statistical problem here involved is this: How can we measure the value, in terms of prospective sales, of coupon (or other) responses from a given medium of advertising, in the absence of direct information on percentage of conversions and value of conversions? We shall make the problem more definite by assuming that the medium in question is a magazine, and we shall refer to the advertiser as the A company. Suppose that another company, the B company, which sells direct by mail, has used this same medium of advertising and has found that sales conversions are high enough to make its use profitable. In other words, the B company rates the quality of coupon responses as excellent. Our problem is this: Is the A company justified in guiding itself by the measured results of the B company, and in concluding that the responses to its own advertising in this magazine are also of high quality?

This question, like so many practical questions involving a generous sprinkling of unknowns, requires qualification before an answer can be even attempted. If we knew nothing of the two companies, and of their products, it would be rash indeed to give an affirmative answer. The B company, for example, may be selling patent medicines direct by mail, while the A company is selling a tooth paste through its dealers. It would be unsound to conclude, without supporting evidence, that because these two companies obtained equal response to their advertising, the potential sales values of their responses are equal. On the other hand, if the A and B companies are selling similar, although not necessarily identical, classes of products, but by different methods, we might feel safe in transferring the conclusions of the B company to the A company, provided that there was no significant difference in the "come-on" features of the two ads. Evidently there is a large element of judgment involved in the answers to such questions. This is as it should be. We are not justified in expecting general answers to questions that clearly demand a careful attention to the particular circumstances of each case.

We may well ask, Where have we got to in all this? Are we any nearer to a solution of these important problems than when we started? I think that we are, if only in the sense that we have better formulated the question, and that we have realized how essentially its answer depends on a detailed study of the individual case and on the quality of the required judgment.

In practice, there is an increasing tendency to rely more and more, in estimating the quality of responses from a medium, on the demonstrated sales results of other advertisers, especially of those who are classed as "respectable" or "acceptable." This tendency has led to an increasingly scientific approach in the problem of advertising and will surely result in increased efficiency in the use of the advertising dollar.

We come now, in our survey of the statistical problems of advertising, to the problem of the general advertiser. He is in the unfortunate position (statistically speaking) of having no direct response of any sort which could guide him in his future selection of mediums. He plants his acres and at harvest time his granaries are full or empty, as the case may be. But he does not know which acres, or which seeds, produced their proportionate share of the harvest. He knows only that if he does not plant, there can be no harvest.

Much of what we said about the coupon advertiser who is ignorant of sales per advertising dollar applies to the general advertiser. But the latter must make his decision to advertise or not to advertise in a given medium without tangible evidence of any sort, in so far as his own advertising results are concerned. In this situation he has three alternatives: He can make a direct judgment of the probable effectiveness of the medium for advertising of the contemplated type, he can rely on tests of the medium carried out by advertisers who are able to check sales, or he can rely on statistical surveys. In practice, the advertising expert relies more often than not, in the absence of reliable and conclusive surveys, on his judgment of the medium, after its sales representatives have been permitted to present their case. But the current tendency, at least of the more progressive experts, is to take account also of the directly measurable "pulling power" of a medium. Especially is this procedure effective when the pulling power of the medium is measured by the results of a diversified list of advertisers.

The advertising effectiveness of a medium, say a magazine, is often estimated in terms of its "reader interest." This idea needs a good deal of clarification and is, at best, not overly satisfactory as a criterion. Reader interest presumably means just what it says, namely interest in the editorial pages of the magazine, as distinguished from the advertising pages. It is ordinarily assumed that if a magazine has reader interest, it follows that its advertising is productive. This conclusion does not in the least follow from the rules of logic, and if it is correct, as it seems to be in general, with certain qualifications, it is an empirical conclusion. The practical importance of correlating reader interest with advertising power is very great for advertisers who are unable to check sales.

But if an advertiser can find out the sales produced by a magazine (for example), he has no possible reason to concern himself with this question of reader interest. He has spent $20,000, let us say, in advertising in a certain magazine, and his records show that he has made a net profit of $5,000. From his point of view it is a matter of indifference whether a single article in the magazine was read. He is sure of one thing; his ad was read. His profit is in the bank, and he will continue this advertising, other things being equal, until it ceases to show a profit.

None of these advertising problems is so simple, in a practical sense, that its answer can be reduced to a formula. Each demands an intelligent and unbiased study of every pertinent fact, and the statistician who fails in such a study is an unmitigated liability.

Statistical surveys can be a fertile source of information as to the effectiveness of an advertising medium, but they require great skill in execution and their proper execution may involve expenses out of proportion to the value of the results. As an example, a national weekly once conducted a survey to determine the number of people who had paid sufficient attention to a page advertisement to remember that they had seen it. Results showed a total of about 20 per cent, based correctly on total circulation, rather than on the fictitious "number of readers."

Before considering the various types of error to which such surveys are subject, let us consider some of the implications of this result as applied to certain advertising problems. If 20 per cent of readers see the page ad, it is equivalent to saying that the chance that a particular reader will see the ad is 1 in 5. Let us apply this conclusion to the much-discussed question of "duplicated circulation." Suppose that a second weekly magazine carrying this same ad reaches this same reader and that the chance that the reader will see it is also 1 in 5. This is an assumption that requires the most careful analysis, but for the purposes of argument we shall let it pass. Then the chance that the reader will see *both* ads is only 1 in 25. This indicates that, even on the assumption that it is largely advertising waste for the reader to see the second ad, the amount of waste involved is small.

What is the chance that the reader will see *at least one* of the two ads? Having considered this matter in Chapter V, we shall not commit the error of adding the two chances, which would give a chance of 2 in 5. The correct result, as we know, is a chance of 9 in 25. But this differs little from 2 in 5, or 10 in 25, and we see that, under the conditions and assumptions outlined, the advertiser loses little, at the worst, by duplicated circulation. If three magazines, instead of two, reach the reader, all carrying the same ad, the corresponding results are as follows: Chance that the reader will see the ad in all three magazines is 1 in 125; chance that he will see it in at least two of the magazines is 13 in 125; chance that he will see it in at least one of the magazines is 61 in 125. If the original survey gives an initial chance of other than 1 in 5, it is easy to revise these conclusions accordingly.

We have implied that such surveys are to be received with

the utmost caution. This means, first of all, that the sample on which their conclusions are based must not be a biased sample, a point discussed in Chapter XIV. In the second place, assuming the correctness of this first step, how can we know whether or not it is justifiable to transfer the results of such a survey from the case of one magazine to that of another? Without attempting an exhaustive list of the requisite precautions, let us list a few of them in the form of questions: Does the result depend on the number of pages in the issue of the magazine tested? Does it depend on the number and prominence of the "competitive" ads in the issue? Does it depend on the editorial or other features of the particular issue in question? Does it depend on peculiarities inherent in the particular class of readers of the particular magazine? Does it depend on seasonal or economic factors?

This series of questions should serve to indicate the danger of accepting too uncritically the results of these surveys.

The majority of the statistical surveys that are furnished the advertiser, with the purpose of showing him how he should spend his advertising budget to the best advantage (of someone) are excellent examples of fallacies in statistics. We can apply to them the remark Goethe once made in another connection: "Their correct conclusions are not new, while their new conclusions are not correct." These surveys are seldom made by professional statisticians. They are usually thrown together by men without technical knowledge and with no understanding of scientific method. The formula used in their construction consists in suppressing every unfavorable fact, and in piling up the favorable ones, until the bombarded prospect, unless he is well fortified against such attacks, is quite overwhelmed with facts and figures.

Perhaps the fairest way to look at advertising solicitations of this sort is to compare them with the tactics of a criminal lawyer defending a client. Legal ethics, while demanding that he tell the truth, do not demand that he tell the whole truth, to the detriment of his case!

In summary, we can arrange the problems of measuring advertising response, considered from the statistical point of view, in the following order of increasing complexity: Coupon or direct response advertising, to be followed by *direct-mail* selling or a direct-sale offer in the original ad; coupon or direct-response advertising, to be followed by *indirect* selling (through dealers, for example); publicity or name advertising—general advertising. It is not only an interesting problem, but one of the most important practical problems in modern business, to discover statistical means of measuring the value (in dollars of net profit per dollar of advertising) of the second and third of these methods, particularly the third.

The field of advertising is a broad one. It includes the psychological problems of appeal to eye and ear, of typographical and pictorial appeal, of the subtleties of color combinations. It includes the art of salesmanship, for advertising is "salesmanship in print" (perhaps, to take account of such modern developments as radio and television, this dictum should be revised to "salesmanship by remote control"). It includes many problems in the solution of which statistics can play no part. On the other hand, quite aside from the clear-cut applications of statistics which we are discussing, there are other problems which permit a statistical attack, in addition to the more obvious approach. As an example, the effectiveness of color in advertising can be estimated either by use of facts discovered by the psychologist in his laboratory, or by the direct, experimental, and statistical method.

When it is possible to answer a question by sound statistical procedure, the answer so obtained can be relied on to an extent seldom possible in the case of answers otherwise obtained. This is merely an affirmation of the supremacy of the experimental method, when that method is applicable. If you want to know the color of John Smith's eyes, you will do better to look at them than to base your conclusion on the color of Smith eyes over the last six or eight generations. Similarly, if we wish to measure the appeal of an ad, and can possibly contrive a statistical check, we shall be surer in our conclusions than if we base them on theoretical psychological reasoning. We therefore feel justified in "dragging in" the statistical method whenever we can soundly do so.

The technique of conducting experiments in direct-action advertising follows as closely as possible that used in the physical sciences. The ad is prepared in the two forms that are to be compared; let us call them Form 1 and Form 2. Form 1 is presented to one group of prospects, call it A, and, simultaneously, Form 2 is presented to another group, call it B. It is known in advance either that the two groups "pull" the same (on the same ad), or that one of them pulls more than the other by a determined amount. With this information it is easy to determine, from the count of responses to the two ads, whether Form 1 or Form 2 is the better puller. If nothing is known in advance of the groups to which the ads are presented, a second test can be made in which Form 1 is presented to group B, and Form 2 to group A. In a physical experiment involving weighing, this corresponds to reversing the scalepans of a chemical balance. If one of the forms pulled better on the first test because it went to the more productive of the two groups, the tables will be reversed on the second test, and this fact will be evident. This latter method is slow and is subject to the further objection that the two groups may respond differently to a repeat ad, so that it is rarely used.

This experimental technique may be clearer from an example. Several years ago one of the smaller mail-order houses decided to make some tests with a view to saving money in the printing of its catalog, which had always been printed in two colors, with the exception of the cover, which was in four colors. They therefore printed several hundred thousand copies of one edition in one-color rotogravure, the four-color cover remaining. The printing cost per catalog in rotogravure was considerably *less*. The test was conducted as follows: A large number of towns were selected from various regions of the country, approximately in proportion to the distribution of the mailing list, and the customers' names from these towns divided alphabetically into two equal parts. The two-color catalog was mailed to one half, the rotogravure catalog to the other, the mailings for a given town taking place on the same day. Orders from the two catalogs were distinguished by code letters before the catalog number of each item listed. When these orders were tabulated, it was found that the rotogravure catalog had outpulled the two-color catalog by more than 15 per cent; in other words, for every 100 orders from the two-color catalog there were more than 115 from the rotogravure one. Furthermore, when the total amounts of the orders were compared, the difference was even more marked; in other words, the average order produced by the rotogravure edition was slightly larger. This point was of importance, for an increase in the number of orders, accompanied by even a mild decrease in the dollars value of the average, would not necessarily mean an increase in net profits.

The conclusion was, then—assuming there was no error in the test—overwhelmingly in favor of adopting a rotogravure catalog, which would bring an initial saving in the cost of printing the catalog and a large profit from the increased orders.

Now it rarely happens that economies in business lead to an absolute increase in sales volume, and this result was very naturally viewed with suspicion by the management of the mail-order house. Owing to an additional feature of the test, however, it was possible to apply a very satisfactory check, as follows: In this concern it was customary to tabulate the orders by states, on account of certain significant differences from one geographical region to another. In the test the orders were therefore listed by states, making it possible to compare the pull of the two catalogs in more than twenty states. It was found that in each of these twenty regions the rotogravure catalog produced more orders than its rival. The increase was by no means the same in different regions, however; it varied from as high as 22 per cent to as low as 6 per cent, according to the geographical situation of the region. The high increases came from the Southern states, the low increases from the Northern states, with intermediate increases corresponding to intermediate states.

This possibility of dividing the test into more than twenty smaller tests, each of which gave the same qualitative answer, plus the fact that the variations in quantitative results were plainly correlated with geographical situation, constituted an extremely satisfactory verification of the correctness of the indications. The probability that they were due to an error was reduced practically to zero.

As a matter of fact, the indicated conclusions were adopted by the concern, and subsequent catalogs were printed in rotogravure. As an additional precaution it was decided to repeat the test on a slightly smaller scale, but this time, in view of the high probability that the first test was correct, it was done the other way about. The main run of catalogs was in rotogravure; the test edition of some two hundred thousand was in two colors. The second test gave quantitative results very close to those of the first test; the practical conclusions were identical.

In the experiment just described, the factor of economic changes in the groups of customers during the tests was eliminated by splitting the towns into two arbitrary, equal lists, and by making the two mailings simultaneously. Any change affecting the one group was certain to affect the other in the same way, for the two groups were statistically identical. A great advantage of this method of conducting the tests is that a comparison of the relative pulling powers of the catalogs (or ads) is possible as soon as a sufficient number of orders (or responses) has been received. The sole advantage in waiting to draw conclusions until approximately all forthcoming responses have been received is that a greater number of responses is then available, and the probability that a significant difference in the figures is due merely to chance fluctuations is correspondingly decreased.

This method of mailed tests to split town lists, which avoids so many of the serious pitfalls in testing, is possible only when there is no conflict whatever between the two pieces of advertising involved. In the case of the two-color versus rotogravure printing test, there was clearly no conflict. More generally, the same is true of tests concerning such matters as wording or arrangement of the ad, size of the ad, illustrations used, or whether or not a coupon should be added. On the other hand, if one ad read "Send 25 cents for a 1-ounce sample," and the other read "Send 35 cents for a 1-ounce sample," the method would be sheer folly. It would be essential that the groups receiving these two ads be far enough apart to avoid odious comparisons. In general, when there is any change in the offer affecting price or terms, split-town tests cannot be used.

The results of this same test raised an interesting question. Why was the rotogravure catalog so much more effective, as compared to the two-color catalog, in Southern states than in Northern states? This is a question that could be answered, on a statistical basis, only by extensive surveys, and such surveys are expensive. As the answer was of no practical value to the mail-order house, no expenditure for surveys was advisable and none was made. But it costs nothing to invent a theoretical explanation, provided that no verification is required. Perhaps Northern people responded less to rotogravure printing because they were more accustomed to this form of printing, in newspaper supplements, for example, and part of the novelty had worn off. We have mentioned this last problem merely to illustrate the contrast between the statistical approach and other approaches. The former yields either no answer or one to

302 ·

which a definite probability can be attached. The latter yields an answer in which the element of judgment is very large, and very difficult to express as a probability.

Let us now confine ourselves entirely to cases where the responses to the advertising can be traced to their source, and where they are promptly received. We then have a statistical situation that has some striking resemblances to firing a rifle at a target, as discussed in Chapter XV, and some striking differences. In both cases the small causes that influence the results are beyond the range of analysis. In the case of advertising, these small causes are the influences that determine whether or not the individual prospect will respond to the ad. To attempt to discover them would be like setting out on the track of John Doe, who committed an unknown crime on an unknown day at an unknown hour. What the advertiser does know is that a certain advertising distribution reaches certain classes of people, meaning collections of people who have one or more significant features in common, such as living in a small town, or being interested in gardening; or he knows, or thinks he knows, that each of a certain list of prospects owns a banjo. This is information of a statistical character.

In the problem of the target we saw that the shot pattern follows a definite statistical law, due to the fact that the small accidental errors that cause the spreading show no preference for one direction rather than another. In this advertising problem we are looking for the same sort of statistical order, and there is in fact such an order, but there is nothing tangible to correspond to the accidental errors. There are only the completely inaccessible factors that influence the individual prospect in his decision to respond or not to respond. It would be absurd to attempt to say that these influences are as likely to be in one "direction" as another. Thus the basis of the previous statistical solution is missing in this case, and we are forced to appeal at once to the court of experience. We shall then find that the distribution of the responses in time (i.e., by days, weeks, or months; as the case may be), which corresponds to the distribution of the shots on the target in the other problem, is *not* symmetrical about any center. The advertising problem contains extra degrees of complication.

If we were in possession of a statistical solution of the problem similar to that for the shot patterns, it would give us an idea in advance of the distribution of the responses in time, by days or by weeks, so that the amount of actual experience necessary to make successful predictions of the total number of responses from a relatively small number of initial ones would be greatly decreased. Having no such theoretical solution, it is necessary to build up the equivalent by actual testing, that is, by advertising and counting responses. When this has once been done for a given medium of advertising (it might be a magazine or a mailed catalog), it will serve afterward as a basis for prediction, always provided that it is used under nearly the same conditions as those under which it was obtained. As our experience increases, we learn how to allow for changes in these conditions, different rates of distribution of the advertising matter, for instance, and in the end have acquired a smooth-working machine that enables us to predict with considerable accuracy the total number of responses to be received over a period of two or three months from the number actually received during a short initial period of a week or two. In the individual case the accuracy of the prediction will depend to a large extent on the number of responses involved, the larger the scale of operations, the greater the accuracy to be expected.

One of the principal problems is to learn how much, or how little, importance is to be attached to results based on a small number of responses. This is best done in terms of chances or probabilities. Suppose, for example, that a new style of ad is being tested in comparison with another style, previously in use, and that under the same conditions the old ad produces 200 responses, the new one 250. The numbers involved are small, and any conclusions based on them correspondingly uncertain. In practice, if it is possible, the test is repeated several times, until the responses to the two ads become numerous enough to remove most of the uncertainty. But it frequently happens that the necessity for immediate action, along whatever lines are indicated by the small test, prevents this repetition, and it becomes a matter of some importance to determine the reliability of the results of the first test.

The problem can be expressed as follows: What is the chance that the increase of fifty responses, in the new ad, is due merely to a chance fluctuation and not to the fact that the new ad pulls better than the old? Let us see how such a problem is solved in practice. Imagine that we have before us the results of many tests, say one hundred, in which the old ad, or a similar one, was used. The total number of responses received will vary from one test to another. We can list these numbers in the form of a frequency distribution, just as we have done in other cases in previous chapters. This frequency distribution will have an average; suppose that it comes out to be 218. The problem is then to study the scattering of the values of the table about the value 218, and to do this the first step is to compute the standard deviation, as has been pointed out so many times in these pages. Finally, we obtain the probability that the number of responses received on a test will differ from the average, 218, by more than a certain number. It might turn out, for instance, that there is 1 chance in 2 that the number of responses will differ from 218 by more than 16. There is 1 chance in 2, then, that the number of responses will be between 202 and 234, inclusive. Mathematical theory then tells us, as we saw in Chapter XV, Table XXV, that there is 1 chance in 5.65 that the number of responses will fall between 186 and 250.

These results give us the answer to our problem. We see that the old ad, with 200 responses, differed from the average by 18; we should expect such a deviation on nearly half of a long series of tests; it is in no way remarkable. The test of the new ad, on the other hand, differs from the old test average by 32 responses. There is only 1 chance in 5.65 that such a result would occur, on the assumption that the new ad has the same pulling power as the old. We can say, then, that the probability that the new ad is a better puller than the old is 4.65/5.65, or roughly 4 chances out of 5.

If the cost of the new ad is the same as that of the old, the betting odds are altogether in favor of adopting it, and no one would hesitate to do so. When the cost of the new ad is greater, the above result provides an arithmetical basis for a judgment. It is to be noted that, taking account of the results of the hundred previous tests on the old ad, we cannot assert that the increase in responses on the new ad is 50, as the small test showed. For we now know that the most probable single result on the old ad is 218 responses, so that we must take the increase as 250 minus 218, or 32 responses, an increase of nearly 15 per cent. It may be objected that in the case of the new ad we have used for the most probable number of responses a figure obtained in a single small test. That is true, and it is also true that the probability that this figure would turn out to be the most probable value, if the test on the new ad could be repeated a large number of times, is very small. Nevertheless, we are justified, in the absence of further statistics, in adopting it as the most probable value, *for the chance that the true value is greater than the adopted one is the same as the chance that it is less.*

Failure to appreciate the soundness of this procedure is equivalent to a failure to understand the true character of an advertising problem like the present one. The reader who has followed our attempt to trace the workings of chance through the various fields of endeavor where it crops up, from simple games of chance to advertising and business, will at once appreciate this fact.

In discussing any practical statistical problem, like the one just undertaken, it is always necessary to append a long list of precautionary qualifications. It is easier to take these precautions, in practice, than it is to describe them adequately. In the case of the test of the new ad against the old, we referred to a series of one hundred tests on the old ad. Theoretically these tests should involve the same ad and the same group of prospects. In practice this is never the case; the series of tests must be selected from the available data and adjusted so as to be as nearly comparable as possible. This process involves a large element of statistical judgment, and the reliability of the results depends above all on the quality of this judgment.

In a brief chapter on so extensive a subject as advertising, it is impossible to give anything approaching a comprehensive survey of the role of chance and statistics in advertising problems. We have, perhaps, succeeded in pointing out that these problems are essentially statistical in character, a fact most strangely overlooked in many of the elaborate books devoted to advertising. The majority of their authors, indeed, do not seem to know that there exists a mathematical theory of statistics, designed expressly to handle problems of the very sort encountered in many fields of advertising, and few approach the subject from that experimental point of view which is so largely responsible for the progress of physical science during the last three centuries. The experimental approach is not directly applicable, as we have seen, to some forms of publicity advertising, but it is often indirectly applicable; and in other forms of advertising it should reign supreme. This point of view is emphasized in E. T. Gundlach's book, *Facts and Fetishes in Advertising,* which contains (Part III and Supplement) a sound and convincing presentation of the case for scientific method in advertising, backed up by a wealth of material taken from the author's long experience as head of an advertising agency. On the other hand, statistical method, with all the powerful aid that it is capable of bringing to the solution of the very problems considered, is left out of account. One finds a great deal of sound thinking in terms of numbers, but little in terms of chances.

There is no sharp line of demarcation between advertising problems and business problems, although it is convenient to separate them as far as possible. In the case of a forecast of sales produced by some form of advertising, for instance, the classification really depends on the purpose for which the forecast is made. If the purpose is to arrive quickly at a decision in regard to future advertising, it should come under the head of advertising. If the purpose is to control purchases of merchandise, it is better classified under the head of business. A discussion of problems of the latter type would therefore belong to the subject matter of the next chapter.

CHAPTER XIX

Business and Statistics

THE board of directors of, let us say, the X company is in solemn session, and not in too good a humor. The sales manager has just finished a rather long-winded explanation of why his sales plan, for which a substantial appropriation was voted some months earlier, was not a success. He attributes its failure to a combination of unfavorable circumstances beyond his control, expresses the opinion that it is extremely unlikely that so unfavorable a situation would again be met, and in the end recommends a further appropriation for a repetition of the scheme. One of the directors, a man of intense action, who found it difficult to spare the time to attend the meeting, and who has been fumbling restlessly in his watch pocket during this discourse, loses no time in expressing himself. "Your scheme looked good when you brought it in," he says, "and I think we were very generous in the appropriation that we gave you to try it out. But now that it has been put to the test, and we have before us the actual facts of its performance, there is really nothing to discuss. Theories are all right in their place, but they can't go up against hard facts." These remarks meet with instant approval, a fortunate circumstance from the point of view of everyone except the sales manager; for the speaker is not the only one who has urgent business elsewhere. The meeting is adjourned, the sales plan abandoned.

Of course it is not for us to pass judgment on the scene we have just witnessed, for our knowledge of the affairs of the X company is very probably even less than that of its directors.

The sales manager may be right in attributing the failure of his plan to unusual external conditions, or he may be wrong. But we can say that, although the director-spokesman may well be correct in his conclusion not to continue the plan, his reasons for this conclusion are hopelessly inadequate. He has avoided the only real point of the issue, namely, the soundness of the manager's explanation of the failure and has taken refuge in a fallacious generality.

The element of chance in business operations of this sort is very large, and especially where chance enters, one swallow does not make a summer. A single trial of a plan, subject to the uncertainties of chance events, proves neither its merit nor its lack of merit. If someone were foolish enough to offer you odds of 60 to 40 on tossing heads (it might be a patron of a gambling casino) and you lost the first toss, or even the first several tosses, you would not conclude that the "hard facts" of experience had demonstrated the unsoundness of your bet. On the contrary, you would continue the game unruffled; as the odds are with you, you are certain to win in the long run.

Similarly in the case of the X company. The correct analysis of their problem must involve a conclusion as to the probable odds in favor of, or against, their sales plan. Suppose that the odds *are* in favor of this plan; then it would be as foolish to discontinue it on the basis of one failure as it would be to drop a favorable bet because of one loss. There is one necessary qualification, however, to all such assertions. As we saw in Chapter VIII, the amount risked in testing the plan should be small compared with the total resources of the company, and this factor should be taken account of in computing the odds. It is on this rock that so many businesses founder.

On the other hand, if the plan was properly executed, and if no supporting evidence can be found for the theory that external and uncontrollable influences were responsible for its failure, the practical conclusion must be that the odds are against it and it must be dropped. The chance of an event depends on the amount of pertinent information that we possess concerning the event. In complicated matters, we seldom have at our disposal all the pertinent facts; hence our predictions based on estimates of chances cannot be accurate. Our predictions based on vague, nonnumerical judgments are very much less accurate. Nevertheless, in practical affairs action is essential. We cannot remain inactive in the hope that sometime, somehow, there will be more light on the difficult passages. We take the best available indications and plunge ahead.

The X company's handling of its sales plan is an obvious instance of a type of thinking that is altogether too common in business. It consists in leaving out of account, either wholly or partly, the element of chance that is almost always present in business affairs. The result is a misunderstanding of the proper significance of "facts" and a failure to grade them as to crudeness or refinement. Under the general heading of facts we find such diversified statements as the following:

Customer John Smith ordered 100 dollars' worth of goods today.
Last month the average order was 50 cents higher than for the same month last year.
Last year the total production of automobiles in the United States was 2,500,000.
At present more than 80 per cent of the automobiles on the road are seven or more years old.
Last week it was announced that the Berlin crisis will be referred to the Security Council of the United Nations.
Last week, due to the failure to settle the Berlin crisis outside the United Nations, stock prices on the New York Stock Exchange declined 5 per cent.

The first of each of these pairs of statements represents a crude fact, a simple statement of a past condition, which admits of ready verification or refutation. The second statement of each pair represents a fact, or set of facts, which have been more or less refined. They are structures of facts, built with the purpose of expressing, either directly or by implication, probabilities or facts of still greater refinement. The fact, for instance, that more than 80 per cent of the automobiles on the road are seven or more years old is one way of telling us that there will be a heavy demand for new automobiles for several years to come, a statement with evident further implications for the automobile manufacturers and for our economy as a whole. It is also one way of pointing out one of the effects of a great war.

We can say in general that the cruder the fact, the more we can rely on it and the less we can learn from it. The more refined facts, on the other hand, are pregnant with implications, but these implications are to be received with reasonable caution. Facts have to do with the past, while implications (or at least those we are most interested in) have to do with the future and are usually best stated in terms of chance. When this element of chance is omitted, as it was in our statement above concerning the effect of the Berlin crisis decision on the stock market, the resulting assertion is more dogmatic than reasonable.

One object of statistical study is to convert crude facts into refined facts and to draw from the latter implications as to the future course of events.

Modern business, with its wide ramifications and its extensive complications, provides a fertile field for the science of statistics. In these pages we have repeatedly pointed out that when large numbers of facts, in complex patterns, are to be dealt with, statistics should be employed. It is in fact, as we have said, almost the sole working tool in the social sciences; it plays a highly important role in every science that is far enough along the road to maturity to be able to deal with quantity, that is to say with numbers. To attack more or less complicated problems that can be stated largely in terms of numbers, whether they belong to biology, economics, govern-ment, or business, without availing oneself of the resources of the science of statistics, is much like entering a modern battle armed with a bow and arrow.

Behind the statistical method, and as a matter of fact includ-ing it, is the scientific method itself. When business research is properly executed, it draws freely from whatever field of science seems appropriate to the problem in hand. It may be the experimental techniques of the laboratory, the analytical techniques of mathematical science, or the techniques of sta-tistics that we have been describing. Operations research, re-ferred to in Chapter XVII, is the application of these tech-niques to the broad problems of war; business research is their application to the broad problems of business.

It would be out of place to discuss here this broader concept of business research, except to the extent that it is inextricably tied up with statistical research. Most experiments in business, for instance, are designed for statistical handling and interpre-tation. The opinion poll is one example, for if properly con-ducted an opinion poll is very definitely an experiment. Sim-ilarly, most mathematical analyses in business are based on a substructure of statistics, and their conclusions are therefore inescapably statistical. In all these activities the ideas of the theory of probability are basic. With this understanding we shall as far as possible confine our discussion to the statistical phases of business research.

A great many of the facts with which business deals are numerical; for example, sales, expenses, profits, accounts re-ceivable, surplus, dividends. In fact, everything that can be ex-pressed in dollars is numerical, and so is everything that can be counted, such as number of sales, of customers, of employees, of hours worked, of accidents, and so on. In order to under-stand what is taking place in a business it is essential to keep records of quite a number of things that are not entered on the books of account. For this purpose most large businesses have what they call statistical departments; their activities might be called *statistical bookkeeping*. All these numerical data of every sort are the raw materials for statistical research.

But there are also many facts with which business deals that are not numerical and are therefore not a part of these raw materials. And in this latter category are many of the most important things in business, such matters as judgments of people, style sense, good taste, qualities of leadership, customer relations, public relations, labor relations, and many others. The point we wish to emphasize is that statistical method has very slight application to the problems just cited, so that there exist large and important areas in business in which scientific method, essentially a development of the last three hundred years, has little to offer. Here the judgment of the businessman reigns almost as undisputed (and as unsupported) as it did in the time of the Caesars. It is therefore false to say that the introduction of modern scientific techniques eliminates the need for judgment or intuition. On the contrary, by analyzing and clarifying those complexes of facts to which it is applicable, it tends to sharpen judgment. It provides for it a firmer foundation from which to take off, and it removes from the field of guesswork many problems where guessing and speculation are as out of place as they are in determining the time of sunrise tomorrow.

The possibility of important or even revolutionary improvements in operations, as a result of the use of statistical research methods, has not been entirely overlooked by business and industry, especially the larger and more progressive concerns. This is particularly true of the large manufacturers. General Motors and the Bell Telephone System, to mention but two, carry on extensive statistical research, parallel to and distinct from engineering and general scientific research. The American Radiator Company was a pioneer in taking full advantage of statistical analyses. Many other outstanding examples could be cited. Unfortunately the same cannot be said for the businesses occupied with the distribution of goods. The majority of large retailers and wholesalers have not attempted to set up competent research departments. A very small number have done so with success, but the majority have selected as research directors men with no scientific training or knowledge, most of whom do not know the difference between statistics and arithmetic. Under these conditions a research department is of value only if it avoids technical problems and confines itself to helping the executives of the business in their routine problems, using routine methods. That such departments are sometimes successful merely proves how great is the need for research in modern business. That need can be fully met only by competent, scientifically trained men.

It is not difficult to understand why manufacturers would be the first to see and to utilize the advantages to be gained by applying modern statistical methods to the large-scale problems of their organizations. For most manufacturers depend for survival either directly or indirectly on basic scientific and engineering research, above all the larger concerns. Their managers and executives are brought up, so to speak, on the scientific method, and they have acquired at first hand a respect for its incisiveness and power. They know that, given a problem amenable to scientific techniques, there is no other intelligent way to tackle it. Now the statistical method is the only intelligent approach to an entire group of problems, including many that come under the head of organization and administration. What could be more sensible than for them to adopt it?

Distribution businesses, on the other hand, have no direct contact with basic research, with minor exceptions. The owners and executives do not therefore have the opportunity to observe for themselves the power of the scientific method. To them science is something that is carried on in laboratories and other far-off places—it is something that could not possibly have a useful application to their own business here and now. They incline to the view that the introduction of statistical research techniques would be tantamount to abandoning judgment and common sense in favor of a set of rule-of-thumb procedures, and they are wise enough to know that such a course would be disastrous. If introducing modern statistical methods were in fact a substitute for judgment, this view would be entirely correct. But we have carefully pointed out that intelligently conducted statistics is not a substitute for judgment; on the contrary, it is an important aid to sounder and keener judgment.

In some businesses a blind worship of traditional methods of operating is a serious stumbling block to the introduction of new techniques of any kind. The typical large department store, for example, tends to suffer somewhat from this malady. It was founded two or three generations back by a self-made merchant whose success was based on ability and, more often than not, outstanding character. Usually he had one or two major ideas, around which he built his organization.* There was a natural tendency to worship these ideas, since they were at the root of a great success, and a corresponding tendency for them to freeze and stiffen to the point that they could not evolve with the changing times. Such an idea cannot be copyrighted or patented, and as soon as its public appeal becomes evident, it is widely copied or imitated. The interlopers, however, do not have the same reverent attitude toward the idea and are more likely to modify it to suit current conditions. The situation usually becomes worse when the founder retires and hands over his store-empire to the next generation. At this point there is danger that not only one or two ideas, but a large number, become frozen or partly frozen. In this way reactionary traditions are formed.

Taking business and industry as a whole, there is today rel-

---

* An outstanding merchant once told me that he attributed his great success more to one simple policy than to any other single thing. He was the first merchant in his particular city to announce and carry out a policy of refunding on any purchase, no questions asked. Today this policy, with minor modifications, is too general to attract attention.

atively little first-class statistical research being done. Many competent authorities would have predicted otherwise twenty years ago. During these years and the immediately preceding decades great progress was made in the theory of statistics, but the practical application of these ideas seems to be exceptionally slow, in a fast-moving age. In view of his own experience in the field, the present writer remains convinced that ultimately statistical techniques will be widely adopted by business and industry, to their great benefit. The need is there and the momentum of scientific achievement is behind the development. The magazine *Nature* had this to say in January, 1926:

A large amount of work has been done in developing statistical methods on the scientific side, and it is natural for any one interested in science to hope that all this work may be utilized in commerce and industry. There are signs that such a movement has started, and it would be unfortunate indeed if those responsible in practical affairs fail to take advantage of the improved statistical machinery now available.

Writing in 1924 on the applications of statistics to economics and business, Professor Mills of Columbia says:*

The last decade has witnessed a remarkable stimulation of interest in quantitative methods in business and in the social sciences. The day when intuition was the chief basis of business judgment and unsupported hypothesis the mode in social studies seems to have passed.

Although these hopes have not fully materialized, they have been supported and renewed as a result of some of the scientific work done during the recent war. In Chapter XVII we discussed briefly the development of operations research in England and in this country. At the end of the war the scientists who had carried on this work realized that the same techniques so successfully applied to essential wartime problems are

---

* *Statistical Methods*, Henry Holt & Co., Inc., New York, page ix.

directly applicable to a wide variety of peacetime problems, including many in the field of business. In an article* previously referred to, Dr. J. Steinhardt says:

> The techniques and principles of analysis developed by the group . . . have wide application to modern government and industry. Briefly, these techniques are those of the competent scientist, applied to a large-scale human operation as a whole, with the aim of fitting the operation to its purpose, and of measuring the effectiveness with which the operation is being carried out.

Several of the scientists involved in operations research have expressed the opinion that a systematic application of its methods to business would yield results of the greatest importance. On the other hand, since some of them have not had direct experience with big business and the atmosphere in which it functions, they perhaps underestimate the practical difficulties that intervene. It is too bad that there are such difficulties, especially as it is business itself, and therefore our economy, that are impeded by them, but it would be foolish to gloss them over.

I think it worth while to devote a few pages to an analysis of these difficulties. They involve adjustments and compromises which must be effected if statistics and, more generally, business research are to achieve any reasonable fraction of their potential accomplishments in the field of business. Such a result is all the more important in that, for the first time, there now exists a sizable group of competent scientists, apart from specialists in mathematical statistics and economics, who by direct experience are convinced of the importance and fertility of applying scientific methods to practical affairs.

The first and greatest obstacle, it must be frankly stated, is the attitude of some businessmen, in particular those who consider themselves exponents of sound, conservative principles

* See footnote, page 281.

that have been long tested, the longer the better. This reaction-ary attitude is too often accompanied by other views which are so very reactionary that they are completely unsound; for the progressive-minded have no monopoly whatever on unsound views. Too often a businessman of the old school regards the application of scientific methods to the operating problems confronting him as an unwarranted and ineffective invasion of his private domain, in effect a slur on his abilities. He believes that the traditional methods are the only ones needed, that the methods under discussion here are impractical and visionary. In this matter he justifies himself, or attempts to, by pointing to the difference between "theory" and "practice." What he is really doing, although he would be the last to admit it, is run-ning down his own profession and himself with it. He is saying, in effect, that business is so crude and undeveloped an occupa-tion that, unlike any one of the sciences, or agriculture, or gov-ernment, it cannot profit from the powerful modern statistical techniques that have made such notable contributions in more enlightened fields. Now this is gross calumny. Business occu-pies itself with the practical art of dealing with men, money, machines and materials. Business is private government, as Beardsley Ruml puts it. It can perform its functions well or badly. But its possibilities for development, refinement, and progress are almost unlimited. As long as progress continues in the correlated arts, in the component parts of which business is made, there is every reason to expect business itself to con-tinue to progress at a corresponding rate. And it will do so if the men in charge of its destiny have the necessary breadth of vision to adopt and incorporate whatever in cognate fields is applicable to its refinement and progress, rather than under-mining it from within.

In addition to the born reactionary, there are other obstruc-tive types in business which are developed and nurtured by certain features of the business environment. These men have allowed themselves to be swept off their feet by an overwhelming passion for action which allows them no rest or peace until something is done about something. This is a direct reflection, although an unnecessary one, of the pace that many American businesses set for themselves. Whether or not such a pace is in any way justified is not under discussion here. We are merely pointing out what it does to certain individuals who allow themselves to fall in with the insidious spirit of rushing from one job to another without pausing long enough to reflect on what they are doing. As long as these jobs are a part of a well-worked-out general plan of operations, these men are accomplishing a great deal and are known as the dynamos of the organization. But when success comes to them and they reach the organizational level of planning the actions of others rather than executing directly themselves, they frequently become the unconscious enemies of real progress. They seem to feel that they have reached a plateau of wisdom and knowledge from which it is their function to radiate a perpetual flow of advice and admonition to the lesser echelons, without the necessity for replenishing their stock of information or for reorienting their views in a changing world. They seem to feel that they have passed the point where it is necessary to make a real effort to learn. If a report is more than a certain number of pages, regardless of the complexity of the matter dealt with, it must be condensed. If it is tough reading, it must be simplified. If the simplification is not simple enough, the matter can be delegated to an assistant, who will present a verbal digest that can be absorbed with very little lost time and effort. In this way decisions can be reached rapidly and passed along. Thus their passion for action is gratified, not action as a part of a carefully worked out and coordinated whole, but action for its own sake.

The behavior and attitude of these particular types of businessmen are so at variance with the spirit of science and the scientist that the two are not compatible. No working scientist, no matter how eminent, ever gets to the point where he feels that he can radiate knowledge without perpetual hard work and study. No scientist would think of shirking the reading of a report or paper, no matter how long or how difficult, if it lies in the direction of his interests.

These contrasts may help to make clear the very real difficulty in attempting to bring businessmen of these types into fruitful contact with statistical science and the scientists who must execute it.

If all businessmen were made in these molds, the immediate prospects for statistical applications to business would be very dim indeed. But businessmen, like any other group of people, are highly diversified. Among them you will also find men of unusual breadth and insight, with an excellent understanding of the spirit of science, progressive, forward-looking men who welcome new ideas from other fields that will help business to do its work more efficiently. When a man of this sort dominates a business, in the sense of permeating it with his ideas and attitudes, there is a great opportunity for sound statistical science to show its true value. Here it can function in a spirit of tolerance and cooperation, with every opportunity to demonstrate its capacities by the tangible results it produces. That such conditions can exist I would like to bear witness. During my years in business research I was fortunate enough to find myself on two occasions a member of an organization of this type. But such conditions are relatively rare, as a careful survey readily discloses, and in all fairness it must be stated that by and large the attitude of businessmen toward the introduction of statistical research on anything approaching an appropriate level is not a completely friendly one.

There is a second difficulty concerning the prospects for sound applications of statistics to business, and this one has nothing to do with the attitude of the businessman. For it statistics itself must take the blame. Up to this writing (1948) it has for the most part failed to put its best foot forward. It has allowed itself to become honeycombed with incompetents, from outright quacks at one extreme, to technically trained statisticians without an ounce of common sense or practical intuition at the other. In other fields where quackery could easily run rampant and throw discredit on everyone, from medicine to accounting, organized safeguards have been established. In medicine, where the health of the community is at stake, protection has extended as far as malpractice laws. Medical degrees from universities indicate competence to the general public. Similarly in accounting, the designation of certified public accountant (C.P.A.) exists under state law, and is a useful index of competence. In statistics no parallel situation exists. Any person, no matter how uneducated or even illiterate, can with impunity set himself up as a statistician and, if he can find someone foolish enough to employ him, proceed to practice. When the educational requirements necessary for competence in modern statistics are taken into account, this situation is little short of scandalous. For these requirements are of the same order as those for a medical or legal degree. The simple fact is that the majority of those who call themselves statisticians are quite incapable of reading and understanding a modern statistical text.* It would seem appropriate that the major statistical societies of the country give some attention to this entire matter.

Behind this failure of statistics to organize and protect itself is the failure, on the part of our educational system in general and our universities in particular, to appreciate the importance of its practical applications. We look in vain, for example, in the Harvard Committee's report on education (*General Education in a Free Society*) for a passage indicating that consideration was given to this field. The same is true even in the leading graduate schools of business administration. If the latter fully appreciated the importance of impressing on the prospective businessman the support that he can expect from practical re-

---

* For example, Cramér's *Mathematical Methods of Statistics,* or Wilks's *Mathematical Statistics,* both published by the Princeton University Press.

search, they would require more thorough grounding in the basic ideas of statistics, and they would offer courses on the history of science and the development of scientific method. In such courses the future executive would obtain some understanding of this method and would learn something of its major achievements.* He would then be better equipped to co-operate intelligently with trained scientists in broadening the field of business research.

The net result of this condition is that, by and large, business statistics is not an academically recognized field. Younger men, attracted to the study of statistics, cannot well be attracted to this phase of it when they do not know of its existence. It follows that you will find a relative concentration of competent statisticians in economics, in government, in agriculture, in fact almost everywhere except in the applications to business, where there are relatively few. A contrary trend has recently appeared in some of our universities, however. New and extended courses in statistics are being offered at Princeton, for example. Columbia has established what amounts to a separate department for advanced work in this field. We quote from the *Bulletin of Information* announcing courses for 1947-1948:

A great increase in the use and the teaching of statistics has occurred in the period since the First World War. This development has come about not only because of the fruitful application of established statistical methods in a rapidly increasing number of fields but also because of important discoveries and advances in the theory of statistics which have made available more powerful and accurate statistical methods. . . . Recognizing the growing importance of statistical theory and methods, the Joint Committee on Graduate Instruction decided in May, 1942, to establish for graduate students a program leading to the degree of Doctor of Philosophy in mathematical statistics. . . .

* I have in mind courses along the lines discussed in the recent book by President Conant of Harvard, *On Understanding Science,* Yale University Press, New Haven, Conn., 1947.

On this hopeful note we shall leave the problem of explaining the slow progress of practical applications of statistics and turn to a brief review of some of these applications themselves.

One of the difficulties in describing scientific research is that each problem is, in a sense, a law unto itself. It is usually quite impossible to lay out in advance the tools one will need; for the attack on each problem discloses new and unexpected problems, the solution of which may in the end be of more value than that of the original one. These ever-changing and unexpected developments are in fact one of the fascinations of research and are a perpetual challenge to ingenuity.

In these respects business research does not differ from other varieties. There is only this one guiding and absolutely essential principle: Since business research is wholly practical research which must justify itself in economic terms, it must be wholly guided by practical considerations. The temptation to explore byways which are not promising from this point of view but which may otherwise have great appeal, must be ruthlessly put aside. For an individual with real enthusiasm for research this may not be so easy as it sounds. My advice to business researchers, based on personal experience, is to carry on simultaneously their own private research in some field—outside of business hours. In this way they can painlessly separate the conflicting sides of their natures.

In any case, a basic prerequisite for business research is a strong feeling or instinct for what is practical. This is not at all the same as what is called common sense. Many people have the latter without the former. The four essentials for the competent business statistician are: technical mastery of modern statistics, research ability, thorough knowledge of business, and practical instinct. Without the last one the other three are of no value. I am emphasizing this matter of practical instinct because it is the most difficult of the four qualities to judge and to measure. A business employing a statistician can check on his technical qualifications, if it will go to the trouble; the other is less easy to evaluate.

Let us assume that a business has a properly set up research department, headed by a thoroughly competent director of research. This is not the place for a discussion of the important and interesting organizational problems involved in such a setup. We shall merely take them for granted and assume that the director of research is in a position to operate freely and effectively.* Under these circumstances what does he do?

In a large and complicated business his first difficulty is that there are too many problems urgently demanding solution. One of his first jobs will therefore be to survey the entire structure, in close consultation with management, in order to set up at least a tentative set of priorities. At this point complications and difficulties begin to arise. As previously stated, each problem tackled is likely to lead to one or more new problems, and it is possible that the latter will overshadow the former in importance. This will call for more consultations and more priorities.

Sometimes one of these problems turns out to be so fertile that an entire new branch of operations flows out of its solution. Many years ago, for example, Dr. Shewhart and his associates at the Bell Telephone Laboratories studied the problem of quality control in the manufacturing process, with very important results for industry generally. We shall let Dr. Shewhart describe the work in his own words.†

---

* Dr. Kettering, vice president and head of engineering and scientific research for General Motors, has stated in speeches his personal demands. They might be called the four freedoms of industrial research: (1) Freedom to originate research projects and to decide which are worth while; (2) freedom from pressure to produce a short-term profit; in other words, freedom to take the long-term economic view; (3) freedom in matters of expenditures; (4) freedom from embarrassment—no accountants or auditors allowed on the premises.

† W. A. Shewhart, "Quality Control Charts," *The Bell System Technical Journal*, October, 1926. Also, W. A. Shewhart, "Economic Quality Control of Manufactured Product"; the same for April, 1930.

"What is the problem involved in the control of quality of manufactured product? To answer this question, let us put ourselves in the position of a manufacturer turning out millions of the same kind of thing every year. Whether it be lead pencils, chewing gum, bars of soap, telephones, or automobiles, the problem is much the same. He sets up a standard for the quality of his product and then tries to make all pieces of product conform with this standard. Here his troubles begin. For him standard quality is a bull's-eye, but like a marksman shooting at such a target, he often misses. As is the case in everything we do, unknown or chance causes exert their influence. The problem then is: how much may the quality of a product vary and yet be controlled? In other words, how much variation should we leave to chance?"

The author of these papers has succeeded in setting up statistical methods that enable the production engineer to determine when he has succeeded in sorting out those causes of variation in the manufactured product which should properly, in the interests of economy and efficiency, be eliminated. "That we may better visualize the economic significance of control, we shall now view the production process as a whole. We take as a specific illustration the manufacture of telephone equipment. Picture, if you will, the twenty or more raw materials such as gold, platinum, silver, copper, tin, lead, wool, rubber, silk, and so forth, literally collected from the four corners of the earth and poured into the manufacturing process. The telephone instrument as it emerges at the end of the production process is not so simple as it looks. In it there are 201 parts, and in the line and equipment making possible the connection of one telephone to another, there are approximately 110,000 more parts. The annual production of most of these parts runs into the millions so that the total annual production of parts runs into the billions.

"How shall the production process for such a complicated mechanism be engineered so as to secure the economies of quantity production and at the same time a finished product" whose variations in quality lie within reasonable specified limits? "Obviously, it is often more economical to throw out defective material at some of the initial stages in production rather than to let it pass on to the final stage where it would likely cause the rejection of a finished unit of product." On the other hand, too severe inspections at the early stages in the process would evidently lead to economic loss. "It may be shown theoretically that, by eliminating assignable causes of variability, we arrive at a limit to which it is feasible to go in reducing" the percentage of defective instruments. In addition to reducing the losses due to rejections, these methods reduce the cost of inspection and accomplish other advantages.*

This problem of quality control has close analogies outside the field of manufacturing. Suppose that you have an office operation—it might have to do with handling an index file of customer names—in which considerable precision is required. Of the various types of errors that can be made by the clerks, each involves a certain amount of loss to the business. In such circumstances you have a problem in error control. How extensively and at what point in the process should you check? If the amounts of the losses can be determined or estimated, many of the statistical techniques developed in connection with quality control are applicable here.

But let us get back to the research director, whom we left struggling with an oversupply of statistical problems. Which of them he and the management will select for urgent study depends so largely on the type of business, the current position of the concern, the problems of the industry of which it is a part, and a long list of other things, that generalities on this subject are out of the question. There is one problem, though, to which he is almost certain to come sooner or later, especially

* A technical discussion of the statistics involved is given in *Mathematical Statistics*, by S. S. Wilks, Princeton University Press, Princeton, N. J., 1944, page 220.

during periods of sharply rising costs, or of sharply falling markets, and that is the problem of controlling expenses, as a means to increased profits. Let us see, avoiding details as much as possible, what sort of contribution to this vital problem management can expect from a statistical research department.

We shall begin with some generalities which are nonetheless true for being obvious. Whatever other functions business fulfills, its primary aim, under our economic system, can be expressed as follows: It must maximize its net profits, over the years, and subject to a set of limiting conditions. These conditions are essentially of three sorts: ethical (for example considerations of national or social welfare), legal (for example honoring labor contracts), and those of self-interest (for example, retaining as fully as possible the good will of the public, of suppliers, and so on). In view of the unavoidable vagueness in the definition of this objective, it is clear that a primary and critical function of management is to formulate explicitly the limiting conditions under which profits are to be made, and to steer a course between the extremes of foolish and shortsighted piling up of profits today, at the expense of tomorrow, and needless sacrifice of present profits through overregard for a distant future. In the solution of this problem lies much of the art of management. The practical approach, and the one that wise managements adopt, is to break it down by first studying methods for producing maximum profits for the immediate future, subject only to the more obvious restrictions. Once this problem is skillfully analyzed, the essential long-term conditions are brought in. Short-term profits are then sacrificed to the extent necessary to fulfill them.

Operating under the policies set up in this way, the research department is, or should be, vitally concerned with methods of maximizing net profits.

Disregarding minor technicalities, the net profit of a company is simply income from sales minus expenses. If we are to have the largest possible net profit, we must have the largest possible net sales with the smallest possible expenses. But if sales are increased, so are expenses, since processing additional sales always costs something. The first step to be taken, and the only one we shall discuss here, is to determine this marginal expense, in relation to sales.

Every young man who rents an office and takes the plunge of going into business for himself knows how to tackle this problem; if not, he is not long in business for himself. He takes a sharp pencil and writes down his overhead expenses—rent, light, telephone, stationery, and so on. Then he figures how many orders he must get per week or per month to cover this overhead. That is his break-even point. He is frequently far too optimistic in his estimate of orders, as pointed out in Chapter VIII, and so may ultimately fail through lack of initial capital, but at least he has correctly taken the first step.

This same first step applies to any business, large or small. It consists in breaking down expenses into two categories—those that do not change when sales increase or decrease (fixed or overhead expenses), and those that do change (moving or running expenses). This is merely a crude first approach to these categories. Now in most large businesses this analysis is by no means so simple as it is in many small businesses. We are not referring to the larger numbers that enter the former, but to complexities in principle. If this were not so, these problems could be handled by simple accounting and there would be no need to attack them with the more powerful weapons of mathematical statistics. I shall ask for the indulgence of the reader while I prove this statement, especially as it will be necessary first to make a few more basic but rather obvious remarks.

In the first place, there is no such thing as a fixed expense, strictly speaking. Many of the expenses that remain fixed for moderate fluctuations in sales jump abruptly to new levels if the change in sales is very large. During the growth period of a successful business, for example, its entire structure must be periodically expanded. When it outgrows its quarters it must rent or build new ones. Since it expects to keep on growing, it usually (quite correctly) takes on far more floor space than it could possibly require for some time to come. This means that every one of the many "fixed" expenses that depend on the amount of floor space jump suddenly to a higher level, where they remain fixed unless or until the level of sales changes radically. Similarly, if there is a drastic drop in the sales level, it is sometimes possible to make correspondingly large reductions in certain fixed expenses.

Fixed expenses are therefore subject to sharp change if sales change their level drastically. It goes without saying that they are also subject to change under the impact of other variations, such as a revision of tax rates, improved operating techniques, or changes in general policy.

We come now to the important problem of analyzing moving expenses, those that depend so directly on sales that they vary, or should vary, with every change in sales. How much they vary is what we wish to determine. We can be sure of only one thing in advance, namely that some of them will vary more, in proportion to a given change in sales, than others. It is also obvious that moving expenses, like fixed expenses, will jump to new levels if the fluctuation in sales is large enough to induce a change in the structure of the business. We shall assume that this does not take place. When it does, a fresh study of the problem is usually required.

This definition of moving expense may seem obvious and clear. That it is not so experience attests. In at least one large business that attempted to classify moving and fixed expenses there was an unbelievable amount of confusion. There were two or three conflicting theories abroad in the institution as to how one recognized a moving expense. One theory was that any expense intimately connected with the sales end of the business was per se a moving expense, and its adherents applied it meticulously. In one branch of this institution, to take a small but picturesque illustration, it was customary to serve tea to customers once a week or once a month, I forget which. In any case, attendance had no relation to changes in sales; the cost of the tea did not vary at all. Nevertheless, this infinitesimal expense was carefully segregated and applied to selling expense, which of course was classified as a moving expense.

The correct and the only correct definition of a moving expense is that it moves with sales. The situation is vaguely reminiscent of the Renaissance battles over the Copernican theory, which stated that the earth moves, and of the heresy trial of Galileo. It will be recalled that Galileo left the trial, after his recantation, muttering to himself, "Nevertheless it moves." The same goes for a moving expense.

How do we find out how much a moving expense changes when sales increase or decrease by a few per cent? In very simple cases we see the answer at once, and these simple cases will provide a clue to the statistical approach to the more complicated ones. Here is the simplest case of all. You employ a salesman to sell your product and you pay him 5 per cent of the sales he produces. His compensation, which is an example of a moving expense, is therefore a constant percentage of sales. If we denote it by E, and the sales he produces by S, the relation is simply E equals 5 per cent of S.

Or you may pay your salesman on a different plan. You may give him a fixed salary of fifty dollars per week, say, and in addition 3 per cent of the sales he produces. In this case the relation between expense and sales is: E equals 3 per cent of S plus 50. If we divide both sides of this simple equation by S, we can write it

$$E/S = \text{ratio of expense to sales} = 3/100 + 50/S.$$

Owing to the fixed salary of fifty dollars, the ratio of expense to sales must be greater than 3 per cent. If sales are \$5,000 it is 4 per cent; if they are only \$1,000 it is 8 per cent, and so on.

Elementary as are these considerations, they give us a clue

that turns out to be sound. Perhaps, they tell us, there are more complicated cases also in which a moving expense is made up of two components, one that is fixed in dollars and one that is fixed as a percentage of sales. The latter can be called a *pure* moving expense, since it is directly proportional to sales. The former is obviously what is called *fixed* expense, or overhead. It should be called *pure* overhead, since many of the accounts ordinarily classified as overhead have elements of moving expense in them. On the other hand, practically every account classified as a moving expense has overhead buried in it, and practical experience tells us that this component is almost never small.

In machine production there is the same sort of thing. While the machine is running it turns out a uniform flow of finished pieces. This corresponds to pure moving expense. When the machine is idle for any reason, labor and other expenses continue. They correspond to pure overhead. Sound cost accounting can give these components in terms of pieces and hours, and finally in terms of dollars. This leads to a simple equation for expense in terms of production.

We shall now show how these simple ideas carry over into a more complicated situation, taken from actual experience. A study was made of the weekly salaries of the sales force of a department store. The object was to discover the relation between these total salaries and total sales, provided that there existed a simple and usable statistical relationship. There was no guarantee in advance that this was true. After careful analysis of the numerical data and after the necessary accounting adjustments* it was found by use of the appropriate mathematical techniques that the following extremely simple equation quite accurately gave selling salaries in terms of sales:

$$E = .0270\,S + 8800.$$

* For example, it is preferable to use gross sales rather than net sales, since obviously the former more closely correspond to the productive work of the sales force.

Here E means total sales force salaries for any week in which the sales were S. For example, if S is 400,000, E is 2.70 per cent of 400,000, which is 10,800, plus 8,800, or 19,600. The formula states that for sales of $400,000, selling salaries for the week will be approximately $19,600. The following table* gives the values of E for several values of S, in thousands of dollars, as well as the corresponding percentages of E to S, in other words the percentages of sales force salaries to sales.

TABLE XXXII

S (000)	E (000)	E/S (%)
250	15.6	6.2
275	16.2	5.9
300	16.9	5.6
400	19.6	4.9
600	25.0	4.2

The value of the formula we have just given depends on the accuracy with which it represents the weekly expenses on which it was based. If it represents them so closely that we can feel sure a simple statistical law was at work, then we can accept it as significant. Otherwise it is of no value. For any equation can be "fitted" to any set of data whatever, if no demands are made as to the closeness of the fit. Also the formula adopted should be as simple as possible. The more complicated it is, the less its significance. Our formula actually fitted the data very well. This means that the weekly total of selling salaries, as computed from the formula, using sales as recorded, did not differ very much, on the average, from the actual total selling salaries. Numerically, this difference averaged about 2.5 per cent of the total salaries. This turned out to be a highly satisfactory result, indicating that a simple statistical law was in

* None of the figures given here are the actual figures. However, they are strictly proportional to the latter and therefore serve quite as well for illustrative purposes.

fact operating. The formula was sufficiently accurate to be used, among other things, for practical budgeting purposes.

This account of the procedure is so oversimplified that it may give a quite false impression. Ordinarily there are many difficulties to be overcome. It is not often true, for example, that a moving expense can be expressed so simply in terms of sales. Frequently the work represented by the moving expense is more directly related to other variables of the business, or to combinations of such variables. Number of sales, or number of customers may be more appropriate variables. If the moving expense represents payroll, it is usually better to express it as hours worked. In this way changes in hourly rates of pay are more easily taken account of. Or accounting adjustments of one sort or another may be essential. I recall that a certain large store, knowing that I was working on this problem at the time, sent me their sales and selling salaries by weeks with the request that I determine what relation existed between them. After a great deal of effort and many false starts I was unable to find a simple relation that fitted the data closely enough to be significant. I was, however, confident that some simple type of relation did exist, based on previous experience with a large number of similar studies. I therefore asked the store for more details concerning the accounts making up total selling salaries. To make a long story short, it was finally discovered that certain commission payments to salesmen were being applied to the second week following that in which they were earned. When these payments were segregated and applied to the proper week, a simple formula was found that fitted the data exceptionally well.

Let us return to the formula

$$E = .0270\ S + 8800,$$

and see what we can learn from it. The first thing we observe

is its extreme simplicity. This is in itself a remarkable phenomenon. Here we have a situation involving the weekly salaries of
hundreds of people engaged in making tens of thousands of
individual sales of the most diverse objects, from pins to pianos,
and yet the averaging process works so smoothly from week to
week that this elementary formula sums it all up, with an accuracy of 97.5 per cent. As the weeks pass there are seasonal
shifts in the pattern of sales, leading to considerable fluctuations in the average amount of the sale. Salespeople are not
paid on the same basis throughout the store, and this fact introduces further distortions as the pattern of sales changes sea-
sonally. Above all, perhaps, there is the irregular flow of customers, not only from week to week or from day to day, but
from hour to hour. It would indeed require a bold person to
predict in advance that this apparently confused and irregular
tangle of cause and effect could be expressed by any simple
formula whatever.

The formula states that total selling salaries are composed
of two parts. One is what we have called *pure moving expense*
(directly proportional to sales), the other we have called *pure
overhead* or *fixed expense*. We shall better understand the significance of these component parts if we think in terms of what
a salesclerk actually does throughout a typical day. Her main
job is of course to sell goods to customers. If all her time were
so spent, her entire salary could be classified as pure moving
expense. But this is far from the case. Due to irregularities in
customer traffic there are times when customers are waiting for
her and other times when she is waiting for customers. The
latter represent waste, from the standpoint of production, just
as does the idle time of a machine. The fact that there is no
avoiding a certain amount of this lost time, that it is an essential part of selling, has nothing to do with its proper classification. Nor does the fact that a part of this time can be used in
stockkeeping, which is another part of the clerk's job. For stock

fluctuates only sluggishly with sales, so that at most but a part of such work contributes to pure moving expense. The bulk of it is overhead.

We see from these considerations why there is so large a component of fixed expense in selling salaries and where it comes from. *It does not come from any group of individuals, whose salaries could be segregated, but from each individual on the selling force.* Part of the salary of Susan Jones, salesclerk, belongs to moving expense, the balance to overhead. This is why we said that it is impossible to separate the two by accounting means. Any such effort is doomed to failure.

Likewise doomed to failure, or at least relative failure, is any effort to budget selling salaries from week to week, as sales fluctuate, in terms of percentage of selling salaries to sales. Account must be taken of the overhead component. We see from Table XXXII that when sales are less than about $326,000, more than half of the weekly total of selling salaries is represented by overhead. This is an astonishing fact. This analysis has not led us to a minor adjustment, but to a very major one. As sales fluctuate over their seasonal range, from $250,000 to $600,000, for the period covered by the table, the percentage of selling salaries ranges from 6.2 per cent to 4.2 per cent. (This example is entirely typical.) Any attempt to hold this percentage fixed would be both absurd and disastrous.

We cannot discuss here the various uses of formulas of this type. We shall merely point out that in budgeting they provide an important over-all control of more detailed budgeting. For management, without entering into the details of departmental budgeting, can set an over-all figure within which the sum of the departmental budgets must fall. These over-all figures form a part of management's plan of controlling expenses and profits.

The use of these formulas for budgeting evidently depends on some sort of estimate of future sales. For this purpose, however, the prediction of sales need be made only for one, two, or at the most three weeks in advance. Especially in large and stable businesses this is not an impossible nor even a difficult task. It is accomplished by a study of the seasonal variation of sales, taking account of possible long-term trends, which are usually too small to be significant in so brief a period, and of any special conditions known to apply, an obvious example being holidays. Here again chance plays a role. All such predictions, no matter how skillfully made, are subject to unknown and unforeseeable factors. If the level of sales is sensitive to weather conditions, for instance, predictions will always be somewhat uncertain until such time as meteorology learns how to predict weather accurately for several weeks in advance. But the fact that all such predictions are subject to uncertainties does not destroy their value; for by and large the sales of a business do follow well-defined seasonal patterns. A skillful prediction is therefore a far better guide than no prediction.

In some businesses it is a matter of the utmost importance to be able to estimate demand for a product weeks or even months in advance, with at least fair accuracy. When this can be done, production schedules can be adjusted to the probable level of sales. The first step in studying long-term sales trends is to remove the seasonal variations that are a prominent feature of almost all businesses. There are several ways to do this. Some of the more reliable methods unfortunately are too slow, in the sense that they furnish adjusted data only after several months have elapsed. They are therefore of value only for the analysis of past periods and are largely used in economic studies. It is preferable to select a method that is less accurate but that gives immediate results. After seasonal variation has been removed there remains a set of figures, a line on a chart if you wish, that indicates the long-term growth or shrinkage of the business. In order to predict the future course of this line, it is necessary to go behind the sales themselves and study the customers from whom they come. These customers must be broken down into economic groups and an attempt made to discover a few fundamental indexes of these groups. But that is not enough. These buying powers must be determined weeks or months in advance; otherwise we would have not a sales prediction but a post-mortem. It is clear that these are essentially problems in economics, and difficult ones. Nevertheless, quite accurate sales predictions have been made during relatively stable economic periods, especially in certain manufacturing businesses. In many types of business it would be foolish to make the attempt.

Many of the statistical problems solved by a research department, perhaps even the majority, are special to the particular business and are suggested by concrete situations that arise. In fact, any complicated business situation in which the data are largely numerical is likely to involve at least one such problem. In solving them a wide variety of theoretical and experimental techniques can be employed. By their nature these problems are of interest chiefly in specialized fields and are not suitable for discussion here.

I shall, however, mention one problem that to my knowledge has not been explored and that might have important implications for the retail business generally. The widespread adoption of the five-day, forty-hour week has created operating difficulties in many retail businesses. In most types of business it is feasible to shut up shop one day each week, customarily on Saturday. In this way the entire force is simultaneously on the job five days each week. But it is not possible to close a retail store on Saturday, for this would involve an utter disregard for service to the public. Saturday normally chalks up more sales per hour than any other day in the week, even though, due to population shifts to suburbs and to the effects of the five-day week in other industries, it now represents a smaller proportion of weekly sales than formerly. Under these circumstances it seems natural to propose that retailers close operations on some other day, and from every point of view Monday would be the proper choice. This would obviate the necessity of complicated scheduling whereby ordinarily the only complete day worked by the entire force is Saturday. The inefficiency of this latter arrangement and its unfavorable effect on customer service are too obvious to need comment.

If such a five-day-week plan were adopted, it would mean that all employees would have a two-day week end, instead of two separate days off. As Monday is usually the least busy day of the week, the amount of real inconvenience to the public would be very small. Whatever it amounts to, it is more than offset by the fact that most large stores and innumerable small ones are open at least one evening a week. In addition to the other advantages, the reduced week would result in considerable operating economies for the stores themselves. The amount of such saving could easily be determined by each store.

What, then, are the objections to this plan strong enough to nullify so many obvious advantages? The only major objection that can be adduced is the fear, on the part of retailers, that by being closed one day each week they would lose sales volume. If Monday now carries 14 per cent of weekly sales, they reason, and we close Monday, will we not lose at least a substantial fraction of this volume? If the five-day week were adopted by all but the smallest neighborhood stores, the answer is that sales losses would be negligible. For any sales lost would have to be lost to other industries, and that would represent a change in the buying habits of the American people. It would require a good deal of unsophistication to believe that so major an effect could flow from so minor a cause.

But it is not easy to bring about a change of this sort, requiring widespread cooperation in good faith, especially in an industry that is notably conservative and so keenly sensitive to competition. It is not easy, even though it could be accomplished by any one competitive area, regardless of the balance of the country. Nevertheless, since such a development would materially benefit both stores and employees, without sacrifice of service to the public, it should be done.

Even if it were admitted that the five-day retail week cannot be achieved through cooperative action, it is quite possible that a well-established large store, with a strong customer following, could profitably establish it, regardless of competitors. In order to form a basis for such a judgment a thorough and penetrating research job would be required. On one scale pan is the probable loss of sales—an estimate of this loss would be the focal point of the investigation—on the other the dollars saved through decreased operating expenses, and two intangibles. These intangibles are increased employee good will due to better working schedules, and improved service to the public.

The conclusion might well hinge on management's judgment of the long-term value to the business of these two intangibles. This is a typical situation in business research. After the problem has been successfully analyzed and numerical conclusions have been reached, final action must depend on human judgment, fortified by the knowledge and insight gained in the investigation.

Even the sketchiest outline of the functions of a statistical research department would be incomplete if it did not mention its responsibility to analyze and interpret for the benefit of the management every significant trend of the business and of its environment. This function, if executed with real skill and broad understanding, can be of great importance. In fact, it is not too much to say that a business equipped with such a source of information has a very definite advantage over its competitors, at least if it has an alert and aggressive management, capable of taking full advantage of the added insight into its problems.

It is not the function of the research department to present to management those facts concerning the business that are normally recorded. That is the function of the accounting department, which ordinarily has attached to it a statistical department to record those facts that do not enter the books of account. It is the job of the research department to apply to such facts, and others that come from outside the walls of the business, scientific techniques that throw new light on their significance in terms of the current problems of the business.

In this final chapter we have given our reasons for believing that statistical applications to business are in their infancy and that, inevitably, someday, they will come into their own. How far off that day is we do not venture to guess. We predict only that when it has arrived business will be a more efficiently conducted institution than at present, just as it is more efficient today than it was before the movement to introduce scientific methods began.

It may be said that we are here indulging in propaganda on behalf of statistical science and its practical applications. Quite possibly so. I take it that there are two varieties of propaganda. In the one the object is to bring others to your way of thinking by any available means, true or false, good or bad. In the other the purpose is to present to others the case for ideas in which you believe, without prejudice or distortion, and in a balanced and impartial way. If we are guilty of propaganda, I trust the verdict will be that it is propaganda of the second sort.

APPENDIX I ·

IN CHAPTER II, "Gamblers and Scientists," we referred to the stormy life of Cardan, not in connection with the theory of probability, but as a passing illustration of the fact that the lives of mathematicians were not always as remote and peaceful as they usually are today. It is something of a coincidence that we selected this particular illustration. For in 1953 Oystein Ore, Sterling Professor of Mathematics at Yale, published a scholarly and penetrating book* on the life and work of Cardan (Cardano), in which it appears that this versatile Italian, among his other accomplishments, correctly enunciated some of the fundamental principles of the theory of probability a century in advance of Galileo, Pascal and Fermat. It is therefore necessary to rewrite the history of this subject and give the honors to Cardan as the first pioneer.

This fundamental work was contained in one of the 111 manuscript books left unpublished at his death, and was not published until 1653. Its title is *Liber de Ludo Aleae* (*The Book on Games of Chance*) a translation of which, by Professor S. H. Gould, appears in Professor Ore's book.

It is natural to inquire why the historians of science have so grossly overlooked this first work on probability, over a period of three hundred years. The answer seems to be that the book is obscure and difficult. It is written in a manner that would be unthinkable in modern scientific works. For example, more than once Cardan attacks the same problem several times, using different approaches, until finally he finds the correct train of reasoning. Under such circumstances one would expect an author to revise his work and delete the erroneous sections. But not so Cardan; he leaves untouched his earlier efforts, to the confusion of the reader, much as though he were writing a diary.

Cardano, the Gambling Scholar; Princeton University Press.

A second and more formidable difficulty is that Cardan, breaking new ground in his book, uses terms that are both vague and obscure, and that possess no precise modern equivalents.

To penetrate this jungle, clear it of both poisonous and dead undergrowth, and put it in order, was a task requiring a mathematician with great insight and patience, plus the collaboration of classical scholars to translate the Latin of the sixteenth century. So it is possible to understand why three centuries were to pass before Cardan's pioneering work received its full due. The history of science owes a great debt to Professor Ore for this illuminating work.

Cardan's *The Book on Games of Chance* wanders over the field of gambling, with much advice to the reader as to how and under what circumstances he should play, with warnings against dishonest gamblers, marked cards and loaded dice, and with a discussion of the ethics of gambling. Section 10, for example, is entitled: "Why gambling was condemned by Aristotle." The book is discursive in the extreme, with occasional reminiscences from the author's life, in which gambling played no small part. It is written without any apparent plan, and in order to study Cardan's reasoning about dice and card games, it is necessary to separate the pertinent from the extraneous.

Cardan examines throws of both six-sided dice, like those in current use, and four-sided dice (astragals), in use since antiquity. As to card games, the modern reader is altogether lost, for the game referred to is one called *primero,* long since forgotten. The important thing is that Cardan, in the end, managed to formulate enough sound ideas on the principles of the theory of chance to merit the role newly ascribed to him.

In the 1950 edition of this book we devoted several pages at the beginning of Chapter XVII ("Statistics at Work") to some problems connected with foot races, starting with the history of the mile run. Since that time every record quoted has been broken, including the four-minute mile "barrier," and it seems worth while to bring these figures up to date and to add a few comments.

First we must make the necessary additions to Table XXVIII (page 271) as follows:

TABLE XXVIII A

Year	Athlete	Country	Time
1954	Bannister	England	3:59.4
1954	Landy	Australia	3:58.0
1957	Ibbotson	England	3:57.2
1958	Elliott	Australia	3:54.5

In line with our remarks on page 272 this final period might be called the British Commonwealth period. When it began it is difficult to say, due for one thing to the disqualification of the Swedish champions. It is also worth noting, in this connection, that during and following World War II, in fact from 1937 to 1954, the record was broken only by athletes from non-belligerent countries.

We note that between Hagg's record of 1945 and Bannister's of 1954 there was an improvement of exactly 2 seconds in nine years. Since Landy also broke the record again that year, we can say that the improvement was 3.4 seconds in nine years. But in the succeeding four years the record fell another 4.9 seconds, and nothing in the modern history of the mile run would lead us to expect a drop of this order. Let us see what we can make of it.

Much has been written in the newspapers and magazines about the four-minute-mile "barrier," supposed to exist in the minds of the athletes, and it is this "barrier" that was cracked by Bannister. Anyone who breaks the record for the mile run certainly deserves the highest praise, and if we say that a careful examination of the facts indicates that in all probability this "barrier" existed in the minds of the sports writers, not the athletes, we do not intend to detract in any way from Bannister's performance. We shall shortly approach this problem from a quite different direction. In the meantime we would point out that almost certainly the greatest performance in the history of the mile run was that of Elliott, and his record run came four years after the mile had been run under four minutes. For in 1958 Elliott lowered the record set by Ibbotson the preceding year by the amazing amount of 2.7 seconds. What makes this performance so completely outstanding is the fact that not since 1866, when the official mile run was only one year old, has any one lowered the record by so large an amount in a single year.

Let us look at this question of the four-minute mile a little more closely. On page 272 we refer to a statistical analysis of Table XXVIII, which we made under the simple assumption that the various complex factors which enter the running of a mile race, and which are largely responsible for the many irregularities disclosed by the table, "averaged out sufficiently to permit an over-all conclusion" over the period 1865 to 1945. Our conclusion was that if the same general trend governs the future, the record for the mile could go as low as 3:44. We did not apply this analysis to the question of the four-minute mile because, at the time the work was done (1947), we did not know that this was a matter of interest. It is easy, however, using this same statistical analysis, to "predict," as of 1947, when the four-minute mile would be run. Upon making the calculations we found to our astonishment that the year predicted for the four-minute mile was

1954, the year that Bannister did it. We say that we were astonished by this result because, as pointed out in the text, the analysis as carried out was superficial, in the sense that it did not have behind it an expert knowledge of the subject. This is emphasized on page 272. It is therefore not reasonable to expect such a formula to yield accurate results for the future. In fact, a large deviation from the formula has already occurred, namely Elliott's 1958 mark of 3:54.5. On the basis of our analysis this level should not be reached until the year 1986.* Whether or not this record is a freak, in the sense that it will require one or two decades to lower it substantially, only the future will disclose. Our guess, and we shall attempt to justify it shortly, is that there has been an improvement in the art of running, and that the record will not stand.

It is possible to refine the above work somewhat by a careful study of Table XXVIII, preferably in the form of a graph. Leaving out of account the first four records, there was a relatively poor performance from 1874 to 1895 inclusive, which reached its nadir in 1881. After a gap of sixteen years, there follows a long period, from 1911 to 1942 inclusive, which has a quite uniform trend, but at a higher level than that of the preceding period. From 1943 to 1961, there has been further improvement, with Elliott's dramatic 1958 record capping the period.

*Readers with some mathematical knowledge might be interested to see the exact equation to which this 1947 analysis leads. Let x be the time, measured in years, with the origin ($x = 0$) at the year 1925, so that for a race run in 1865, $x = -60$, while for one run in 1945, $x = 20$. Let y be the time to run one mile, measured in seconds, with the origin ($y = 0$) at 240 secs., equal to 4 mins. Then the formula is

$$y = -16.3 + 23.7(10)^{-.00552x}.$$

As x increases, the second term on the right side of the equation gets smaller and smaller, so that y approaches -16.3, or 223.7 secs., or 3:43.7. To find the predicted date of the four-minute mile, put $y = 0$, and solve the resulting equation for x, using logarithms. This gives $x = 29.6$, or the year 1954.6.

If we work out a new curve (equation) so adjusted as to make the years 1911 to 1942 the standard, so to speak, we obtain a different result. On this basis the ultimate possible record for the mile turns out to be about 3:35 instead of 3:44. In this analysis, as in the previous one, we have taken no account of records set since 1945. It is interesting to see what this new equation* predicts for future records. The results are given in the following table:

TABLE XXXIII

Year	Predicted Record	Actual Record
1954	4:02.3	3:58.0
1957	4:00.4	3:57.2
1958	4:00.2	3:54.5
1970	3:57.3	
1980	3:55.1	
1990	3:53.2	

We see that this new equation is less successful in predicting the date of the four-minute mile, giving 1958.6. As for Elliott's 1958 record it predicts that, based on the trend to and including 1945, it should be reached in 1983. On the basis of either of the equations it is clear that, at least for the 1950's, there is evidence of a new and lower trend.

To make clear that the new records shown in Table XXXIII represent at least a temporarily lowered trend, it will be well to show to what extent the theoretical statistical equation represents the actual records prior to 1954. This is done, for a few typical years, in Table XXXIV, page 349. If the first of the two theoretical equations had been used, a quite comparable table would have resulted. Neither curve

---

*Again readers familiar with some mathematics may wish to see the form of the equation. Using the same conventions as in the previous footnote in this appendix, the equation is

$$y = -25.4 + 35.5(10)^{-.00432x}.$$

fits the actual records well during the early years, when there were considerable irregularities. Between them, however,

TABLE XXXIV

Year	Theoretical Record	Actual Record	Difference (in secs.) (Actual minus Theoretical)
1866	4:38.4	4:39.0	0.6
1868	4:37.2	4:33.2	— 4.0
1884	4:28.0	4:18.4	— 9.6
1913	4:14.6	4:14.4	— 0.2
1931	4:08.0	4:09.2	1.2
1934	4:07.0	4:06.8	— 0.2
1937	4:06.1	4:06.4	0.3
1942	4:04.6	4:04.6	0.0

they indicate that the recent records are at a lower level than the history of the mile run would indicate.

There are many possible explanations for this improved running. One is the fact that the generation now setting records was born at a time when great advances in the science

TABLE XXX A

Distance		Speed in Feet per Second
Yards	Meters	
100		32.3
220		33.0
440		28.9
880		24.7
(1,093.6)	1,000	23.8
(1,640.4)	1,500	22.8
1,760		(one mile) 22.5
(2,187.2)	2,000	21.7
(3,280.8)	3,000	20.8
3,520		(two miles) 20.6
5,280		(three miles) 20.0
(5,468.0)	5,000	20.1

of nutrition had been made, advances which undoubtedly resulted in greater physical stamina as well as an increase in the average height of individuals.

At first sight one might be tempted to believe that there was a psychological basis for these rapidly falling records, based on the fact that the four-minute "barrier" had been broken. If these higher running speeds were confined to the mile run, there might be something to this theory. But this is not the case, as we shall see.

In Table XXX on page 276 we listed the speeds in feet per second, corresponding to the world records as of 1947 for all standard running races from 100 yards to 5000 meters. Since every one of these records has now been broken, we have prepared a revised table, Table XXXA, given on page 349. An examination of this table discloses that there is nothing remarkable about the current record for the mile, as compared to the records for other distances. It appears that Elliott's performance, although an extraordinary improvement for the mile run, is very much in line with other records set during the past decade. If we subject this table to a statistical analysis, of the type used on Table XXVIII, this fact and some not so apparent become evident. In the study* we

---

*The equation of the curve is

$$y = 16.7 + 9.70(10)^{-.00111x}.$$

Here x is the length of the race measured, for convenience, in units of ten yards, so that a distance of 880 yards corresponds to $x = 88$. y is the speed, measured in feet per second. This equation fits the data quite well, with a probable error of 0.3 ft. per sec. This error would be reduced if distances of three miles and over were omitted. It appears that in the longer races, including the six-, ten- and fifteen-mile distances (not shown in the table), the champion runner is able to sustain a speed higher than what one would expect from a study of shorter races. This effect is less pronounced when one comes to the marathon (46,145 yards), but even here the record shows an average speed of 17.1 ft. per sec., while our theoretical equation gives 16.7. Incidentally, if our statistical result were interpreted literally, it would state that a man can run any distance whatever at an average speed of 16.7 ft. per sec., a palpable absurdity closely related to the one given on page 273. Actually, there are no simple mathematical curves, of the required type, that end abruptly and, if there were, one would have no way to know where the curve should terminate. There is a similar situation in life insurance, where the curves used to represent the mortality rates run on to infinity. They

have omitted distances of less than 880 yards, as they belong in a different category. It appears from the study that, relatively speaking, the highest speeds are attained at 880 yards, at three miles, and at 5,000 meters. The poorest relative speed is at 2,000 meters, the second poorest being the mile. However, the differences are slight. So we must conclude that the art of running, as a whole, has made great progress in recent years. It will be interesting to see to what extent this rate of progress can be maintained in the future.

will tell you the probability that a man will die between the ages of 1000 and 1001, provided that he reaches the age of 1000. The tails of these curves are therefore quite meaningless, but it is convenient to use the curves and the tails do no harm. The same sort of thing is true in our case.

Index

INDEX

A CATALOG OF SELECTED
DOVER BOOKS
IN ALL FIELDS OF INTEREST

A CATALOG OF SELECTED DOVER
BOOKS IN ALL FIELDS OF INTEREST

CONCERNING THE SPIRITUAL IN ART, Wassily Kandinsky. Pioneering work by father of abstract art. Thoughts on color theory, nature of art. Analysis of earlier masters. 12 illustrations. 80pp. of text. 5⅜ x 8½. 23411-8 Pa. $4.95

ANIMALS: 1,419 Copyright-Free Illustrations of Mammals, Birds, Fish, Insects, etc., Jim Harter (ed.). Clear wood engravings present, in extremely lifelike poses, over 1,000 species of animals. One of the most extensive pictorial sourcebooks of its kind. Captions. Index. 284pp. 9 x 12. 23766-4 Pa. $14.95

CELTIC ART: The Methods of Construction, George Bain. Simple geometric techniques for making Celtic interlacements, spirals, Kells-type initials, animals, humans, etc. Over 500 illustrations. 160pp. 9 x 12. (Available in U.S. only.) 22923-8 Pa. $9.95

AN ATLAS OF ANATOMY FOR ARTISTS, Fritz Schider. Most thorough reference work on art anatomy in the world. Hundreds of illustrations, including selections from works by Vesalius, Leonardo, Goya, Ingres, Michelangelo, others. 593 illustrations. 192pp. 7⅛ x 10¼. 20241-0 Pa. $9.95

CELTIC HAND STROKE-BY-STROKE (Irish Half-Uncial from "The Book of Kells"): An Arthur Baker Calligraphy Manual, Arthur Baker. Complete guide to creating each letter of the alphabet in distinctive Celtic manner. Covers hand position, strokes, pens, inks, paper, more. Illustrated. 48pp. 8¼ x 11. 24336-2 Pa. $3.95

EASY ORIGAMI, John Montroll. Charming collection of 32 projects (hat, cup, pelican, piano, swan, many more) specially designed for the novice origami hobbyist. Clearly illustrated easy-to-follow instructions insure that even beginning papercrafters will achieve successful results. 48pp. 8¼ x 11. 27298-2 Pa. $3.50

THE COMPLETE BOOK OF BIRDHOUSE CONSTRUCTION FOR WOODWORKERS, Scott D. Campbell. Detailed instructions, illustrations, tables. Also data on bird habitat and instinct patterns. Bibliography. 3 tables. 63 illustrations in 15 figures. 48pp. 5¼ x 8½. 24407-5 Pa. $2.50

BLOOMINGDALE'S ILLUSTRATED 1886 CATALOG: Fashions, Dry Goods and Housewares, Bloomingdale Brothers. Famed merchants' extremely rare catalog depicting about 1,700 products: clothing, housewares, firearms, dry goods, jewelry, more. Invaluable for dating, identifying vintage items. Also, copyright-free graphics for artists, designers. Co-published with Henry Ford Museum & Greenfield Village. 160pp. 8¼ x 11. 25780-0 Pa. $10.95

HISTORIC COSTUME IN PICTURES, Braun & Schneider. Over 1,450 costumed figures in clearly detailed engravings–from dawn of civilization to end of 19th century. Captions. Many folk costumes. 256pp. 8⅜ x 11¾. 23150-X Pa. $12.95

CATALOG OF DOVER BOOKS

STICKLEY CRAFTSMAN FURNITURE CATALOGS, Gustav Stickley and L. & J. G. Stickley. Beautiful, functional furniture in two authentic catalogs from 1910. 594 illustrations, including 277 photos, show settles, rockers, armchairs, reclining chairs, bookcases, desks, tables. 183pp. 6½ x 9¼. 23838-5 Pa. $11.95

AMERICAN LOCOMOTIVES IN HISTORIC PHOTOGRAPHS: 1858 to 1949, Ron Ziel (ed.). A rare collection of 126 meticulously detailed official photographs, called "builder portraits," of American locomotives that majestically chronicle the rise of steam locomotive power in America. Introduction. Detailed captions. xi+ 129pp. 9 x 12. 27393-8 Pa. $13.95

AMERICA'S LIGHTHOUSES: An Illustrated History, Francis Ross Holland, Jr. Delightfully written, profusely illustrated fact-filled survey of over 200 American lighthouses since 1716. History, anecdotes, technological advances, more. 240pp. 8 x 10¾. 25576-X Pa. $12.95

TOWARDS A NEW ARCHITECTURE, Le Corbusier. Pioneering manifesto by founder of "International School." Technical and aesthetic theories, views of industry, economics, relation of form to function, "mass-production split" and much more. Profusely illustrated. 320pp. 6⅛ x 9¼. (Available in U.S. only.) 25023-7 Pa. $9.95

HOW THE OTHER HALF LIVES, Jacob Riis. Famous journalistic record, exposing poverty and degradation of New York slums around 1900, by major social reformer. 100 striking and influential photographs. 233pp. 10 x 7⅞. 22012-5 Pa. $11.95

FRUIT KEY AND TWIG KEY TO TREES AND SHRUBS, William M. Harlow. One of the handiest and most widely used identification aids. Fruit key covers 120 deciduous and evergreen species; twig key 160 deciduous species. Easily used. Over 300 photographs. 126pp. 5⅜ x 8½. 20511-8 Pa. $3.95

COMMON BIRD SONGS, Dr. Donald J. Borror. Songs of 60 most common U.S. birds: robins, sparrows, cardinals, bluejays, finches, more–arranged in order of increasing complexity. Up to 9 variations of songs of each species. Cassette and manual 99911-4 $8.95

ORCHIDS AS HOUSE PLANTS, Rebecca Tyson Northen. Grow cattleyas and many other kinds of orchids–in a window, in a case, or under artificial light. 63 illustrations. 148pp. 5⅜ x 8½. 23261-1 Pa. $5.95

MONSTER MAZES, Dave Phillips. Masterful mazes at four levels of difficulty. Avoid deadly perils and evil creatures to find magical treasures. Solutions for all 32 exciting illustrated puzzles. 48pp. 8¼ x 11. 26005-4 Pa. $2.95

MOZART'S DON GIOVANNI (DOVER OPERA LIBRETTO SERIES), Wolfgang Amadeus Mozart. Introduced and translated by Ellen H. Bleiler. Standard Italian libretto, with complete English translation. Convenient and thoroughly portable–an ideal companion for reading along with a recording or the performance itself. Introduction. List of characters. Plot summary. 121pp. 5¼ x 8½. 24944-1 Pa. $3.95

TECHNICAL MANUAL AND DICTIONARY OF CLASSICAL BALLET, Gail Grant. Defines, explains, comments on steps, movements, poses and concepts. 15-page pictorial section. Basic book for student, viewer. 127pp. 5⅜ x 8½. 21843-0 Pa. $4.95

THE CLARINET AND CLARINET PLAYING, David Pino. Lively, comprehensive work features suggestions about technique, musicianship, and musical interpretation, as well as guidelines for teaching, making your own reeds, and preparing for public performance. Includes an intriguing look at clarinet history. "A godsend," *The Clarinet,* Journal of the International Clarinet Society. Appendixes. 7 illus. 320pp. 5⅜ x 8½. 40270-3 Pa. $9.95

HOLLYWOOD GLAMOR PORTRAITS, John Kobal (ed.). 145 photos from 1926-49. Harlow, Gable, Bogart, Bacall; 94 stars in all. Full background on photographers, technical aspects. 160pp. 8⅜ x 11¼. 23352-9 Pa. $12.95

THE ANNOTATED CASEY AT THE BAT: A Collection of Ballads about the Mighty Casey/Third, Revised Edition, Martin Gardner (ed.). Amusing sequels and parodies of one of America's best-loved poems: Casey's Revenge, Why Casey Whiffed, Casey's Sister at the Bat, others. 256pp. 5⅜ x 8½. 28598-7 Pa. $8.95

THE RAVEN AND OTHER FAVORITE POEMS, Edgar Allan Poe. Over 40 of the author's most memorable poems: "The Bells," "Ulalume," "Israfel," "To Helen," "The Conqueror Worm," "Eldorado," "Annabel Lee," many more. Alphabetic lists of titles and first lines. 64pp. 5 3/16 x 8¼. 26685-0 Pa. $1.00

PERSONAL MEMOIRS OF U. S. GRANT, Ulysses Simpson Grant. Intelligent, deeply moving firsthand account of Civil War campaigns, considered by many the finest military memoirs ever written. Includes letters, historic photographs, maps and more. 528pp. 6⅛ x 9¼. 28587-1 Pa. $12.95

ANCIENT EGYPTIAN MATERIALS AND INDUSTRIES, A. Lucas and J. Harris. Fascinating, comprehensive, thoroughly documented text describes this ancient civilization's vast resources and the processes that incorporated them in daily life, including the use of animal products, building materials, cosmetics, perfumes and incense, fibers, glazed ware, glass and its manufacture, materials used in the mummification process, and much more. 544pp. 6⅛ x 9¼. (Available in U.S. only.) 40446-3 Pa. $16.95

RUSSIAN STORIES/PYCCKNE PACCKA3bl: A Dual-Language Book, edited by Gleb Struve. Twelve tales by such masters as Chekhov, Tolstoy, Dostoevsky, Pushkin, others. Excellent word-for-word English translations on facing pages, plus teaching and study aids, Russian/English vocabulary, biographical/critical introductions, more. 416pp. 5⅜ x 8½. 26244-8 Pa. $9.95

PHILADELPHIA THEN AND NOW: 60 Sites Photographed in the Past and Present, Kenneth Finkel and Susan Oyama. Rare photographs of City Hall, Logan Square, Independence Hall, Betsy Ross House, other landmarks juxtaposed with contemporary views. Captures changing face of historic city. Introduction. Captions. 128pp. 8¼ x 11. 25790-8 Pa. $9.95

AIA ARCHITECTURAL GUIDE TO NASSAU AND SUFFOLK COUNTIES, LONG ISLAND, The American Institute of Architects, Long Island Chapter, and the Society for the Preservation of Long Island Antiquities. Comprehensive, well-researched and generously illustrated volume brings to life over three centuries of Long Island's great architectural heritage. More than 240 photographs with authoritative, extensively detailed captions. 176pp. 8¼ x 11. 26946-9 Pa. $14.95

NORTH AMERICAN INDIAN LIFE: Customs and Traditions of 23 Tribes, Elsie Clews Parsons (ed.). 27 fictionalized essays by noted anthropologists examine religion, customs, government, additional facets of life among the Winnebago, Crow, Zuni, Eskimo, other tribes. 480pp. 6⅛ x 9¼. 27377-6 Pa. $10.95

FRANK LLOYD WRIGHT'S DANA HOUSE, Donald Hoffmann. Pictorial essay of residential masterpiece with over 160 interior and exterior photos, plans, elevations, sketches and studies. 128pp. 9¼ x 10¾. 29120-0 Pa. $12.95

THE MALE AND FEMALE FIGURE IN MOTION: 60 Classic Photographic Sequences, Eadweard Muybridge. 60 true-action photographs of men and women walking, running, climbing, bending, turning, etc., reproduced from rare 19th-century masterpiece. vi + 121pp. 9 x 12. 24745-7 Pa. $12.95

1001 QUESTIONS ANSWERED ABOUT THE SEASHORE, N. J. Berrill and Jacquelyn Berrill. Queries answered about dolphins, sea snails, sponges, starfish, fishes, shore birds, many others. Covers appearance, breeding, growth, feeding, much more. 305pp. 5¼ x 8¼. 23366-9 Pa. $9.95

ATTRACTING BIRDS TO YOUR YARD, William J. Weber. Easy-to-follow guide offers advice on how to attract the greatest diversity of birds: birdhouses, feeders, water and waterers, much more. 96pp. 5³⁄₁₆ x 8¼. 28927-3 Pa. $2.50

MEDICINAL AND OTHER USES OF NORTH AMERICAN PLANTS: A Historical Survey with Special Reference to the Eastern Indian Tribes, Charlotte Erichsen-Brown. Chronological historical citations document 500 years of usage of plants, trees, shrubs native to eastern Canada, northeastern U.S. Also complete identifying information. 343 illustrations. 544pp. 6½ x 9¼. 25951-X Pa. $12.95

STORYBOOK MAZES, Dave Phillips. 23 stories and mazes on two-page spreads: Wizard of Oz, Treasure Island, Robin Hood, etc. Solutions. 64pp. 8¼ x 11. 23628-5 Pa. $2.95

AMERICAN NEGRO SONGS: 230 Folk Songs and Spirituals, Religious and Secular, John W. Work. This authoritative study traces the African influences of songs sung and played by black Americans at work, in church, and as entertainment. The author discusses the lyric significance of such songs as "Swing Low, Sweet Chariot," "John Henry," and others and offers the words and music for 230 songs. Bibliography. Index of Song Titles. 272pp. 6½ x 9¼. 40271-1 Pa. $9.95

MOVIE-STAR PORTRAITS OF THE FORTIES, John Kobal (ed.). 163 glamor, studio photos of 106 stars of the 1940s: Rita Hayworth, Ava Gardner, Marlon Brando, Clark Gable, many more. 176pp. 8⅜ x 11¼. 23546-7 Pa. $14.95

BENCHLEY LOST AND FOUND, Robert Benchley. Finest humor from early 30s, about pet peeves, child psychologists, post office and others. Mostly unavailable elsewhere. 73 illustrations by Peter Arno and others. 183pp. 5⅜ x 8½. 22410-4 Pa. $6.95

YEKL and THE IMPORTED BRIDEGROOM AND OTHER STORIES OF YIDDISH NEW YORK, Abraham Cahan. Film Hester Street based on *Yekl* (1896). Novel, other stories among first about Jewish immigrants on N.Y.'s East Side. 240pp. 5⅜ x 8½. 22427-9 Pa. $7.95

SELECTED POEMS, Walt Whitman. Generous sampling from *Leaves of Grass*. Twenty-four poems include "I Hear America Singing," "Song of the Open Road," "I Sing the Body Electric," "When Lilacs Last in the Dooryard Bloom'd," "O Captain! My Captain!"–all reprinted from an authoritative edition. Lists of titles and first lines. 128pp. 5³⁄₁₆ x 8¼. 26878-0 Pa. $1.00

PIANO TUNING, J. Cree Fischer. Clearest, best book for beginner, amateur. Simple repairs, raising dropped notes, tuning by easy method of flattened fifths. No previous skills needed. 4 illustrations. 201pp. 5⅜ x 8½. 23267-0 Pa. $6.95

HINTS TO SINGERS, Lillian Nordica. Selecting the right teacher, developing confidence, overcoming stage fright, and many other important skills receive thoughtful discussion in this indispensible guide, written by a world-famous diva of four decades' experience. 96pp. 5⅜ x 8½. 40094-8 Pa. $4.95

THE COMPLETE NONSENSE OF EDWARD LEAR, Edward Lear. All nonsense limericks, zany alphabets, Owl and Pussycat, songs, nonsense botany, etc., illustrated by Lear. Total of 320pp. 5⅜ x 8½. (AVAILABLE IN U.S. ONLY.) 20167-8 Pa. $7.95

VICTORIAN PARLOUR POETRY: An Annotated Anthology, Michael R. Turner. 117 gems by Longfellow, Tennyson, Browning, many lesser-known poets. "The Village Blacksmith," "Curfew Must Not Ring Tonight," "Only a Baby Small," dozens more, often difficult to find elsewhere. Index of poets, titles, first lines. xxiii + 325pp. 5⅜ x 8¼. 27044-0 Pa. $8.95

DUBLINERS, James Joyce. Fifteen stories offer vivid, tightly focused observations of the lives of Dublin's poorer classes. At least one, "The Dead," is considered a masterpiece. Reprinted complete and unabridged from standard edition. 160pp. 5$\frac{3}{16}$ x 8¼. 26870-5 Pa. $1.00

GREAT WEIRD TALES: 14 Stories by Lovecraft, Blackwood, Machen and Others, S. T. Joshi (ed.). 14 spellbinding tales, including "The Sin Eater," by Fiona McLeod, "The Eye Above the Mantel," by Frank Belknap Long, as well as renowned works by R. H. Barlow, Lord Dunsany, Arthur Machen, W. C. Morrow and eight other masters of the genre. 256pp. 5⅜ x 8½. (Available in U.S. only.) 40436-6 Pa. $8.95

THE BOOK OF THE SACRED MAGIC OF ABRAMELIN THE MAGE, translated by S. MacGregor Mathers. Medieval manuscript of ceremonial magic. Basic document in Aleister Crowley, Golden Dawn groups. 268pp. 5⅜ x 8½. 23211-5 Pa. $9.95

NEW RUSSIAN-ENGLISH AND ENGLISH-RUSSIAN DICTIONARY, M. A. O'Brien. This is a remarkably handy Russian dictionary, containing a surprising amount of information, including over 70,000 entries. 366pp. 4½ x 6⅛. 20208-9 Pa. $10.95

HISTORIC HOMES OF THE AMERICAN PRESIDENTS, Second, Revised Edition, Irvin Haas. A traveler's guide to American Presidential homes, most open to the public, depicting and describing homes occupied by every American President from George Washington to George Bush. With visiting hours, admission charges, travel routes. 175 photographs. Index. 160pp. 8¼ x 11. 26751-2 Pa. $11.95

NEW YORK IN THE FORTIES, Andreas Feininger. 162 brilliant photographs by the well-known photographer, formerly with *Life* magazine. Commuters, shoppers, Times Square at night, much else from city at its peak. Captions by John von Hartz. 181pp. 9¼ x 10¾. 23585-8 Pa. $13.95

INDIAN SIGN LANGUAGE, William Tomkins. Over 525 signs developed by Sioux and other tribes. Written instructions and diagrams. Also 290 pictographs. 111pp. 6⅛ x 9¼. 22029-X Pa. $3.95

ANATOMY: A Complete Guide for Artists, Joseph Sheppard. A master of figure drawing shows artists how to render human anatomy convincingly. Over 460 illustrations. 224pp. 8⅜ x 11¼. 27279-6 Pa. $11.95

MEDIEVAL CALLIGRAPHY: Its History and Technique, Marc Drogin. Spirited history, comprehensive instruction manual covers 13 styles (ca. 4th century through 15th). Excellent photographs; directions for duplicating medieval techniques with modern tools. 224pp. 8⅜ x 11¼. 26142-5 Pa. $12.95

DRIED FLOWERS: How to Prepare Them, Sarah Whitlock and Martha Rankin. Complete instructions on how to use silica gel, meal and borax, perlite aggregate, sand and borax, glycerine and water to create attractive permanent flower arrangements. 12 illustrations. 32pp. 5⅜ x 8½. 21802-3 Pa. $1.00

EASY-TO-MAKE BIRD FEEDERS FOR WOODWORKERS, Scott D. Campbell. Detailed, simple-to-use guide for designing, constructing, caring for and using feeders. Text, illustrations for 12 classic and contemporary designs. 96pp. 5⅜ x 8½. 25847-5 Pa. $3.95

SCOTTISH WONDER TALES FROM MYTH AND LEGEND, Donald A. Mackenzie. 16 lively tales tell of giants rumbling down mountainsides, of a magic wand that turns stone pillars into warriors, of gods and goddesses, evil hags, powerful forces and more. 240pp. 5⅜ x 8½. 29677-6 Pa. $6.95

THE HISTORY OF UNDERCLOTHES, C. Willett Cunnington and Phyllis Cunnington. Fascinating, well-documented survey covering six centuries of English undergarments, enhanced with over 100 illustrations: 12th-century laced-up bodice, footed long drawers (1795), 19th-century bustles, l9th-century corsets for men, Victorian "bust improvers," much more. 272pp. 5⅜ x 8¼. 27124-2 Pa. $9.95

ARTS AND CRAFTS FURNITURE: The Complete Brooks Catalog of 1912, Brooks Manufacturing Co. Photos and detailed descriptions of more than 150 now very collectible furniture designs from the Arts and Crafts movement depict davenports, settees, buffets, desks, tables, chairs, bedsteads, dressers and more, all built of solid, quarter-sawed oak. Invaluable for students and enthusiasts of antiques, Americana and the decorative arts. 80pp. 6½ x 9¼. 27471-3 Pa. $8.95

WILBUR AND ORVILLE: A Biography of the Wright Brothers, Fred Howard. Definitive, crisply written study tells the full story of the brothers' lives and work. A vividly written biography, unparalleled in scope and color, that also captures the spirit of an extraordinary era. 560pp. 6⅛ x 9¼. 40297-5 Pa. $17.95

THE ARTS OF THE SAILOR: Knotting, Splicing and Ropework, Hervey Garrett Smith. Indispensable shipboard reference covers tools, basic knots and useful hitches; handsewing and canvas work, more. Over 100 illustrations. Delightful reading for sea lovers. 256pp. 5⅜ x 8½. 26440-8 Pa. $8.95

FRANK LLOYD WRIGHT'S FALLINGWATER: The House and Its History, Second, Revised Edition, Donald Hoffmann. A total revision—both in text and illustrations—of the standard document on Fallingwater, the boldest, most personal architectural statement of Wright's mature years, updated with valuable new material from the recently opened Frank Lloyd Wright Archives. "Fascinating"—*The New York Times*. 116 illustrations. 128pp. 9¼ x 10¾. 27430-6 Pa. $12.95

PHOTOGRAPHIC SKETCHBOOK OF THE CIVIL WAR, Alexander Gardner. 100 photos taken on field during the Civil War. Famous shots of Manassas Harper's Ferry, Lincoln, Richmond, slave pens, etc. 244pp. 10⅝ x 8¼. 22731-6 Pa. $10.95

FIVE ACRES AND INDEPENDENCE, Maurice G. Kains. Great back-to-the-land classic explains basics of self-sufficient farming. The one book to get. 95 illustrations. 397pp. 5⅜ x 8½. 20974-1 Pa. $7.95

SONGS OF EASTERN BIRDS, Dr. Donald J. Borror. Songs and calls of 60 species most common to eastern U.S.: warblers, woodpeckers, flycatchers, thrushes, larks, many more in high-quality recording. Cassette and manual 99912-2 $9.95

A MODERN HERBAL, Margaret Grieve. Much the fullest, most exact, most useful compilation of herbal material. Gigantic alphabetical encyclopedia, from aconite to zedoary, gives botanical information, medical properties, folklore, economic uses, much else. Indispensable to serious reader. 161 illustrations. 888pp. 6½ x 9¼. 2-vol. set. (Available in U.S. only.) Vol. I: 22798-7 Pa. $9.95
Vol. II: 22799-5 Pa. $9.95

HIDDEN TREASURE MAZE BOOK, Dave Phillips. Solve 34 challenging mazes accompanied by heroic tales of adventure. Evil dragons, people-eating plants, bloodthirsty giants, many more dangerous adversaries lurk at every twist and turn. 34 mazes, stories, solutions. 48pp. 8¼ x 11. 24566-7 Pa. $2.95

LETTERS OF W. A. MOZART, Wolfgang A. Mozart. Remarkable letters show bawdy wit, humor, imagination, musical insights, contemporary musical world; includes some letters from Leopold Mozart. 276pp. 5⅜ x 8½. 22859-2 Pa. $7.95

BASIC PRINCIPLES OF CLASSICAL BALLET, Agrippina Vaganova. Great Russian theoretician, teacher explains methods for teaching classical ballet. 118 illustrations. 175pp. 5⅜ x 8½. 22036-2 Pa. $6.95

THE JUMPING FROG, Mark Twain. Revenge edition. The original story of The Celebrated Jumping Frog of Calaveras County, a hapless French translation, and Twain's hilarious "retranslation" from the French. 12 illustrations. 66pp. 5⅜ x 8½. 22686-7 Pa. $3.95

BEST REMEMBERED POEMS, Martin Gardner (ed.). The 126 poems in this superb collection of 19th- and 20th-century British and American verse range from Shelley's "To a Skylark" to the impassioned "Renascence" of Edna St. Vincent Millay and to Edward Lear's whimsical "The Owl and the Pussycat." 224pp. 5⅜ x 8½. 27165-X Pa. $5.95

COMPLETE SONNETS, William Shakespeare. Over 150 exquisite poems deal with love, friendship, the tyranny of time, beauty's evanescence, death and other themes in language of remarkable power, precision and beauty. Glossary of archaic terms. 80pp. 5³⁄₁₆ x 8¼. 26686-9 Pa. $1.00

BODIES IN A BOOKSHOP, R. T. Campbell. Challenging mystery of blackmail and murder with ingenious plot and superbly drawn characters. In the best tradition of British suspense fiction. 192pp. 5⅜ x 8½. 24720-1 Pa. $6.95

THE WIT AND HUMOR OF OSCAR WILDE, Alvin Redman (ed.). More than 1,000 ripostes, paradoxes, wisecracks: Work is the curse of the drinking classes; I can resist everything except temptation; etc. 258pp. 5⅜ x 8½. 20602-5 Pa. $6.95

SHAKESPEARE LEXICON AND QUOTATION DICTIONARY, Alexander Schmidt. Full definitions, locations, shades of meaning in every word in plays and poems. More than 50,000 exact quotations. 1,485pp. 6½ x 9¼. 2-vol. set.
Vol. 1: 22726-X Pa. $17.95
Vol. 2: 22727-8 Pa. $17.95

SELECTED POEMS, Emily Dickinson. Over 100 best-known, best-loved poems by one of America's foremost poets, reprinted from authoritative early editions. No comparable edition at this price. Index of first lines. 64pp. 5 3/16 x 8¼.
26466-1 Pa. $1.00

THE INSIDIOUS DR. FU-MANCHU, Sax Rohmer. The first of the popular mystery series introduces a pair of English detectives to their archnemesis, the diabolical Dr. Fu-Manchu. Flavorful atmosphere, fast-paced action, and colorful characters enliven this classic of the genre. 208pp. 5 3/16 x 8¼. 29898-1 Pa. $2.00

THE MALLEUS MALEFICARUM OF KRAMER AND SPRENGER, translated by Montague Summers. Full text of most important witchhunter's "bible," used by both Catholics and Protestants. 278pp. 6⅝ x 10. 22802-9 Pa. $12.95

SPANISH STORIES/CUENTOS ESPAÑOLES: A Dual-Language Book, Angel Flores (ed.). Unique format offers 13 great stories in Spanish by Cervantes, Borges, others. Faithful English translations on facing pages. 352pp. 5⅜ x 8½.
25399-6 Pa. $8.95

GARDEN CITY, LONG ISLAND, IN EARLY PHOTOGRAPHS, 1869–1919, Mildred H. Smith. Handsome treasury of 118 vintage pictures, accompanied by carefully researched captions, document the Garden City Hotel fire (1899), the Vanderbilt Cup Race (1908), the first airmail flight departing from the Nassau Boulevard Aerodrome (1911), and much more. 96pp. 8⅞ x 11¾. 40669-5 Pa. $12.95

OLD QUEENS, N.Y., IN EARLY PHOTOGRAPHS, Vincent F. Seyfried and William Asadorian. Over 160 rare photographs of Maspeth, Jamaica, Jackson Heights, and other areas. Vintage views of DeWitt Clinton mansion, 1939 World's Fair and more. Captions. 192pp. 8⅞ x 11. 26358-4 Pa. $12.95

CAPTURED BY THE INDIANS: 15 Firsthand Accounts, 1750-1870, Frederick Drimmer. Astounding true historical accounts of grisly torture, bloody conflicts, relentless pursuits, miraculous escapes and more, by people who lived to tell the tale. 384pp. 5⅜ x 8½. 24901-8 Pa. $8.95

THE WORLD'S GREAT SPEECHES (Fourth Enlarged Edition), Lewis Copeland, Lawrence W. Lamm, and Stephen J. McKenna. Nearly 300 speeches provide public speakers with a wealth of updated quotes and inspiration–from Pericles' funeral oration and William Jennings Bryan's "Cross of Gold Speech" to Malcolm X's powerful words on the Black Revolution and Earl of Spenser's tribute to his sister, Diana, Princess of Wales. 944pp. 5⅜ x 8⅜. 40903-1 Pa. $15.95

THE BOOK OF THE SWORD, Sir Richard F. Burton. Great Victorian scholar/adventurer's eloquent, erudite history of the "queen of weapons"–from prehistory to early Roman Empire. Evolution and development of early swords, variations (sabre, broadsword, cutlass, scimitar, etc.), much more. 336pp. 6⅛ x 9¼.
25434-8 Pa. $9.95

AUTOBIOGRAPHY: The Story of My Experiments with Truth, Mohandas K. Gandhi. Boyhood, legal studies, purification, the growth of the Satyagraha (nonviolent protest) movement. Critical, inspiring work of the man responsible for the freedom of India. 480pp. 5⅜ x 8½. (Available in U.S. only.) 24593-4 Pa. $8.95

CELTIC MYTHS AND LEGENDS, T. W. Rolleston. Masterful retelling of Irish and Welsh stories and tales. Cuchulain, King Arthur, Deirdre, the Grail, many more. First paperback edition. 58 full-page illustrations. 512pp. 5⅜ x 8½. 26507-2 Pa. $9.95

THE PRINCIPLES OF PSYCHOLOGY, William James. Famous long course complete, unabridged. Stream of thought, time perception, memory, experimental methods; great work decades ahead of its time. 94 figures. 1,391pp. 5⅜ x 8½. 2-vol. set.
Vol. I: 20381-6 Pa. $14.95
Vol. II: 20382-4 Pa. $14.95

THE WORLD AS WILL AND REPRESENTATION, Arthur Schopenhauer. Definitive English translation of Schopenhauer's life work, correcting more than 1,000 errors, omissions in earlier translations. Translated by E. F. J. Payne. Total of 1,269pp. 5⅜ x 8½. 2-vol. set.
Vol. 1: 21761-2 Pa. $12.95
Vol. 2: 21762-0 Pa. $12.95

MAGIC AND MYSTERY IN TIBET, Madame Alexandra David-Neel. Experiences among lamas, magicians, sages, sorcerers, Bonpa wizards. A true psychic discovery. 32 illustrations. 321pp. 5⅜ x 8½. (Available in U.S. only.) 22682-4 Pa. $9.95

THE EGYPTIAN BOOK OF THE DEAD, E. A. Wallis Budge. Complete reproduction of Ani's papyrus, finest ever found. Full hieroglyphic text, interlinear transliteration, word-for-word translation, smooth translation. 533pp. 6½ x 9¼.
21866-X Pa. $12.95

MATHEMATICS FOR THE NONMATHEMATICIAN, Morris Kline. Detailed, college-level treatment of mathematics in cultural and historical context, with numerous exercises. Recommended Reading Lists. Tables. Numerous figures. 641pp. 5⅜ x 8½.
24823-2 Pa. $11.95

PROBABILISTIC METHODS IN THE THEORY OF STRUCTURES, Isaac Elishakoff. Well-written introduction covers the elements of the theory of probability from two or more random variables, the reliability of such multivariable structures, the theory of random function, Monte Carlo methods of treating problems incapable of exact solution, and more. Examples. 502pp. 5³/₈ x 8¹/₂. 40691-1 Pa. $16.95

THE RIME OF THE ANCIENT MARINER, Gustave Doré, S. T. Coleridge. Doré's finest work; 34 plates capture moods, subtleties of poem. Flawless full-size reproductions printed on facing pages with authoritative text of poem. "Beautiful. Simply beautiful."–*Publisher's Weekly.* 77pp. 9¼ x 12. 22305-1 Pa. $7.95

NORTH AMERICAN INDIAN DESIGNS FOR ARTISTS AND CRAFTSPEOPLE, Eva Wilson. Over 360 authentic copyright-free designs adapted from Navajo blankets, Hopi pottery, Sioux buffalo hides, more. Geometrics, symbolic figures, plant and animal motifs, etc. 128pp. 8⅜ x 11. (Not for sale in the United Kingdom.) 25341-4 Pa. $9.95

SCULPTURE: Principles and Practice, Louis Slobodkin. Step-by-step approach to clay, plaster, metals, stone; classical and modern. 253 drawings, photos. 255pp. 8⅛ x 11.
22960-2 Pa. $11.95

THE INFLUENCE OF SEA POWER UPON HISTORY, 1660–1783, A. T. Mahan. Influential classic of naval history and tactics still used as text in war colleges. First paperback edition. 4 maps. 24 battle plans. 640pp. 5⅜ x 8½. 25509-3 Pa. $14.95

THE STORY OF THE TITANIC AS TOLD BY ITS SURVIVORS, Jack Winocour (ed.). What it was really like. Panic, despair, shocking inefficiency, and a little heroism. More thrilling than any fictional account. 26 illustrations. 320pp. 5⅜ x 8½. 20610-6 Pa. $8.95

FAIRY AND FOLK TALES OF THE IRISH PEASANTRY, William Butler Yeats (ed.). Treasury of 64 tales from the twilight world of Celtic myth and legend: "The Soul Cages," "The Kildare Pooka," "King O'Toole and his Goose," many more. Introduction and Notes by W. B. Yeats. 352pp. 5⅜ x 8½. 26941-8 Pa. $8.95

BUDDHIST MAHAYANA TEXTS, E. B. Cowell and others (eds.). Superb, accurate translations of basic documents in Mahayana Buddhism, highly important in history of religions. The Buddha-karita of Asvaghosha, Larger Sukhavativyuha, more. 448pp. 5⅜ x 8½. 25552-2 Pa. $12.95

ONE TWO THREE . . . INFINITY: Facts and Speculations of Science, George Gamow. Great physicist's fascinating, readable overview of contemporary science: number theory, relativity, fourth dimension, entropy, genes, atomic structure, much more. 128 illustrations. Index. 352pp. 5⅜ x 8½. 25664-2 Pa. $9.95

EXPERIMENTATION AND MEASUREMENT, W. J. Youden. Introductory manual explains laws of measurement in simple terms and offers tips for achieving accuracy and minimizing errors. Mathematics of measurement, use of instruments, experimenting with machines. 1994 edition. Foreword. Preface. Introduction. Epilogue. Selected Readings. Glossary. Index. Tables and figures. 128pp. 5³/₈ x 8¹/₂. 40451-X Pa. $6.95

DALÍ ON MODERN ART: The Cuckolds of Antiquated Modern Art, Salvador Dalí. Influential painter skewers modern art and its practitioners. Outrageous evaluations of Picasso, Cézanne, Turner, more. 15 renderings of paintings discussed. 44 calligraphic decorations by Dalí. 96pp. 5⅜ x 8½. (Available in U.S. only.) 29220-7 Pa. $5.95

ANTIQUE PLAYING CARDS: A Pictorial History, Henry René D'Allemagne. Over 900 elaborate, decorative images from rare playing cards (14th–20th centuries): Bacchus, death, dancing dogs, hunting scenes, royal coats of arms, players cheating, much more. 96pp. 9¼ x 12¼. 29265-7 Pa. $12.95

MAKING FURNITURE MASTERPIECES: 30 Projects with Measured Drawings, Franklin H. Gottshall. Step-by-step instructions, illustrations for constructing handsome, useful pieces, among them a Sheraton desk, Chippendale chair, Spanish desk, Queen Anne table and a William and Mary dressing mirror. 224pp. 8⅛ x 11¼. 29338-6 Pa. $13.95

THE FOSSIL BOOK: A Record of Prehistoric Life, Patricia V. Rich et al. Profusely illustrated definitive guide covers everything from single-celled organisms and dinosaurs to birds and mammals and the interplay between climate and man. Over 1,500 illustrations. 760pp. 7½ x 10⅛. 29371-8 Pa. $29.95

Prices subject to change without notice.

Available at your book dealer or write for free catalog to Dept. GI, Dover Publications, Inc., 31 East 2nd St., Mineola, N.Y. 11501. Dover publishes more than 500 books each year on science, elementary and advanced mathematics, biology, music, art, literary history, social sciences and other areas.